Pinpoint Math

Student Booklet
Level F

Volume 3
Add, Subtract, Multiply, and Divide

Photo Credits
©Andreas Guskos/Shutterstock Inc., cover *background*; ©Shebanov Alexandr/Shutterstock Inc., cover *top image*.

Acknowledgements
Content Consultant:

Linda Proudfit, Ph.D.

After earning a B.A. and M.A in Mathematics from the University of Northern Iowa, Linda Proudfit taught junior- and senior-high mathematics in Iowa. Following this, she earned a Ph.D. in Mathematics Education from Indiana University. She currently is Coordinator of Elementary Education and Professor of Mathematics Education at Governors State University in University Park, IL.

Dr. Proudfit has made numerous presentations at professional meetings at the local, state, and national levels. Her main research interests are problem solving and algebraic thinking.

www.WrightGroup.com

Copyright © 2009 by Wright Group/McGraw-Hill.

All rights reserved. Except as permitted under the United States Copyright Act, no part of this publication may be reproduced or distributed in any form or by any means, or stored in a database or retrieval system, without the prior written permission from the publisher, unless otherwise indicated.

Printed in USA.

Send all inquiries to:
Wright Group/McGraw-Hill
P.O. Box 812960
Chicago, IL 60681

ISBN 978-1-40-456809-9
MHID 1-40-456809-3

1 2 3 4 5 6 7 8 9 MAZ 13 12 11 10 09 08

The McGraw-Hill Companies

Contents

Tutorial Chart viii

Volume 3 Add, Subtract, Multiply, and Divide

Topic 6 Use Addition or Subtraction to Compute

Topic 6 Introduction 1
Lesson 6-1 Add or Subtract Numbers..................................2–4
Lesson 6-2 Add or Subtract Mentally 5–7
Lesson 6-3 Add or Subtract with Estimation 8–10
Lesson 6-4 Add 2-Digit Numbers......................................11–13
Lesson 6-5 Subtract 2-Digit Numbers.................................14–16
Lesson 6-6 Add or Subtract 3-Digit Numbers 17–19
Lesson 6-7 Multidigit Numbers.......................................20–22
Topic 6 Summary 23
Topic 6 Mixed Review 24

Topic 7 Use Multiplication to Compute

Topic 7 Introduction 25
Lesson 7-1 Multiply Multiples of 10 26–28
Lesson 7-2 Estimate Products 29–31
Lesson 7-3 Multiply Multidigit Numbers 32–34
Lesson 7-4 Choose a Method...35–37
Topic 7 Summary 38
Topic 7 Mixed Review 39

Topic 8 Use Division to Compute

Topic 8 Introduction 40
Lesson 8-1 Divide Multiples of 10 41–43
Lesson 8-2 Estimate Quotients..44–46
Lesson 8-3 Divide by 1-Digit Numbers 47–49
Lesson 8-4 Choose a Method..50–52
Lesson 8-5 Multiply and Divide......................................53–55

Volume 3   iii   Level F

Topic 8 Summary 56

Topic 8 Mixed Review 57

Topic 9 Basic Properties of Algebra

Topic 9 Introduction 58

Lesson 9-1 Properties of Addition 59–61

Lesson 9-2 Properties of Multiplication 62–64

Lesson 9-3 The Distributive Property 65–67

Topic 9 Summary 68

Topic 9 Mixed Review 69

Topic 10 Expressions and Equations

Topic 10 Introduction 70

Lesson 10-1 Rules for Patterns 71–73

Lesson 10-2 Write Expressions 74–76

Lesson 10-3 Write Equations 77–79

Lesson 10-4 Solve Equations 80–82

Topic 10 Summary 83

Topic 10 Mixed Review 84

Glossary 85

Word Bank 86

Index 88

Volume 3: Add, Subtract, Multiply, and Divide

Topic 6 Use Addition or Subtraction to Compute

Lesson	Objective	Pages
Topic 6 Introduction	6.2 Use mental arithmetic to find the sum or difference of two 2-digit numbers. 6.4 Add two 2-digit numbers with and without regrouping. 6.5 Subtract two 2-digit numbers with and without regrouping. 6.6 Find the sum or difference of two whole numbers up to three digits long.	1
Lesson 6-1 Add or Subtract Numbers	6.1 Solve addition and subtraction problems with 1- and 2-digit numbers.	2–4
Lesson 6-2 Add or Subtract Mentally	6.2 Use mental math to find the sum or difference of two 2-digit numbers.	5–7
Lesson 6-3 Add or Subtract with Estimation	6.3 Estimate sums and differences of 2-digit numbers.	8–10
Lesson 6-4 Add 2-Digit Numbers	6.4 Add two 2-digit numbers with and without regrouping.	11–13
Lesson 6-5 Subtract 2-Digit Numbers	6.5 Subtract two 2-digit numbers with and without regrouping.	14–16
Lesson 6-6 Add or Subtract 3-Digit Numbers	6.6 Find the sum or difference of two whole numbers up to three digits long.	17–19
Lesson 6-7 Multidigit Numbers	6.7 Use standard algorithms to add and subtract multidigit numbers.	20–22
Topic 6 Summary	Review skills related to addition and subtraction.	23
Topic 6 Mixed Review	Maintain concepts and skills.	24

Topic 7 Use Multiplication to Compute

Lesson	Objective	Pages
Topic 7 Introduction	7.1 Solve problems involving multiplying multiples of 10, 100, and 1,000. 7.2 Estimate products by rounding factors and using mental math techniques. 7.3 Multiply multidigit numbers by 1-digit numbers. 7.4 Solve problems by choosing between using mental math, paper and pencil, or estimation.	25
Lesson 7-1 Multiply Multiples of 10	7.1 Solve problems involving multiplying multiples of 10.	26–28

Volume 3　　　　　v　　　　　Level F

Lesson 7-2 Estimate Products	7.2 Estimate products by rounding factors and using mental math techniques.	29–31
Lesson 7-3 Multiply Multidigit Numbers	7.3 Multiply multidigit numbers by 1-digit numbers.	32–34
Lesson 7-4 Choose a Method	7.4 Solve problems by choosing between using mental math, paper and pencil, or estimation.	35–37
Topic 7 Summary	Review skills related to multiplication.	38
Topic 7 Mixed Review	Maintain concepts and skills.	39

Topic 8 Use Division to Compute

Lesson	Objective	Pages
Topic 8 Introduction	8.1 Solve problems involving dividing multiples of 10. 8.2 Estimate quotients by rounding numbers and using mental math techniques. 8.3 Divide multidigit numbers by 1-digit numbers with and without remainders. 8.5 Solve problems by choosing between using mental math, paper and pencil, or estimation.	40
Lesson 8-1 Divide Multiples of 10	8.1 Solve problems involving dividing multiples of 10.	41–43
Lesson 8-2 Estimate Quotients	8.2 Estimate quotients by rounding numbers and using mental math techniques.	44–46
Lesson 8-3 Divide by 1-Digit Numbers	8.3 Divide multidigit numbers by 1-digit numbers with and without remainders.	47–49
Lesson 8-4 Choose a Method	8.4 Solve problems by choosing between using mental math, paper and pencil, or estimation.	50–52
Lesson 8-5 Multiply and Divide	8.5 Solve word problems by using either multiplication or division.	53–55
Topic 8 Summary	Review skills related to division.	56
Topic 8 Mixed Review	Maintain concepts and skills.	57

Topic 9 Basic Properties of Algebra

Lesson	Objective	Pages
Topic 9 Introduction	9.1 Use the commutative and associative properties of addition. 9.2 Recognize and use the commutative and associative properties of multiplication. 9.3 Know and use the distributive property.	58
Lesson 9-1 Properties of Addition	9.1 Use the commutative and associative properties of addition.	59–61
Lesson 9-2 Properties of Multiplication	9.2 Recognize and use the commutative and associative properties of multiplication.	62–64
Lesson 9-3 The Distributive Property	9.3 Know and use the distributive property.	65–67

Lesson	Objective	Pages
Topic 9 Summary	Review the commutative, associative, and distributive properties.	68
Topic 9 Mixed Review	Maintain concepts and skills.	69

Topic 10 Expressions and Equations

Lesson	Objective	Pages
Topic 10 Introduction	**10.1** Record the rule for a pattern as an expression. **10.2** Write expressions for situations that include an unknown quantity. **10.3** Write equations for word problems that include an unknown quantity. **10.4** Solve simple equations for word problems that include an unknown quantity.	70
Lesson 10-1 Rules for Patterns	**10.1** Record the rule for a pattern as an expression.	71–73
Lesson 10-2 Write Expressions	**10.2** Write expressions for situations that include an unknown quantity.	74–76
Lesson 10-3 Write Equations	**10.3** Write equations for word problems that include an unknown quantity.	77–79
Lesson 10-4 Solve Equations	**10.4** Solve simple equations for word problems that include an unknown quantity.	80–82
Topic 10 Summary	Review writing and solving expressions and equations.	83
Topic 10 Mixed Review	Maintain concepts and skills.	84

Tutorial Guide

Each of the standards listed below has at least one animated tutorial for students to use with the lesson that matches the objective. If you are using the electronic components of *Pinpoint Math*, you will find a complete listing of Tutorial codes and titles when you access them either online or via CD-ROM.

Level F

Standards by topic	Tutorial codes
Volume 3 Add, Subtract, Multiply, and Divide	
Topic 6 Use Addition or Subtraction to Compute	
6.1 Solve addition and subtraction problems with 1- and 2-digit numbers.	6a Using Models to Solve Word Problems
6.3 Estimate sums and differences of 2-digit numbers.	6b Using the Partial Sums Algorithm to Add
6.3 Estimate sums and differences of 2-digit numbers.	6c Using the Same-Change Rule to Subtract
6.4 Add two 2-digit numbers with and without regrouping.	6b Using the Partial Sums Algorithm to Add
6.4 Add two 2-digit numbers with and without regrouping.	6d Solving Word Problems, Example A
6.4 Add two 2-digit numbers with and without regrouping.	6a Using Models to Solve Word Problems
6.5 Subtract two 2-digit numbers with and without regrouping.	6c Using the Same-Change Rule to Subtract
6.5 Subtract two 2-digit numbers with and without regrouping.	6a Using Models to Solve Word Problems
6.6 Find the sum or difference of two whole numbers up to three digits long.	6e Using the Standard Addition Algorithm, Example A
6.6 Find the sum or difference of two whole numbers up to three digits long.	6f Using the Standard Subtraction Algorithm, Example A
6.7 Use standard algorithms to add and subtract multidigit numbers.	6i Using the Standard Addition Algorithm, Example B
6.7 Use standard algorithms to add and subtract multidigit numbers.	6j Using the Standard Subtraction Algorithm, Example B
Topic 7 Use Multiplication to Compute	
7.1 Solve problems involving multiplying multiples of 10.	7a Using the Partial-Products Method to Multiply
7.1 Solve problems involving multiplying multiples of 10.	7b Using Multiples of 10, 100, and 1,000 to Multiply and Divide
7.2 Estimate products by rounding factors and using mental math techniques.	7a Using the Partial-Products Method to Multiply
7.2 Estimate products by rounding factors and using mental math techniques.	7c Using the Standard Multiplication Algorithm
7.3 Multiply multidigit numbers by 1-digit numbers.	7d Solving Word Problems, Example A
7.3 Multiply multidigit numbers by 1-digit numbers.	7a Using the Partial-Products Method to Multiply
7.4 Solve problems by choosing between using mental math, paper and pencil, or estimation.	7e Choosing a Method to Solve Multiplication and Division Problems
Topic 8 Use Division to Compute	
8.1 Solve problems involving dividing multiples of 10.	8a Using Multiples of 10, 100, and 1,000 to Multiply and Divide
8.2 Estimate quotients by rounding numbers and using mental math techniques.	8b Estimating Quotients by Rounding Numbers
8.3 Divide multidigit numbers by 1-digit numbers with and without remainders.	8c Modeling Division
8.3 Divide multidigit numbers by 1-digit numbers with and without remainders.	8d Solving Word Problems, Example C

8.3 Divide multidigit numbers by 1-digit numbers with and without remainders.	8e Using a Standard Long Division Algorithm, Example A
8.4 Solve problems by choosing between using mental math, paper and pencil, or estimation.	8f Using a Standard Long Division Algorithm, Example B
8.4 Solve problems by choosing between using mental math, paper and pencil, or estimation.	8g Using a Standard Long Division Algorithm, Example C
8.5 Solve word problems by using either multiplication or division.	8i Solving Word Problems, Example B
8.5 Solve word problems by using either multiplication or division.	8d Solving Word Problems, Example C
8.5 Solve word problems by using either multiplication or division.	8j Using Multiplication to Check Division
Topic 9 Basic Properties of Algebra	
9.1 Use the commutative and associative properties of addition.	9a Using the Commutative and Associative Properties of Addition
9.2 Recognize and use the commutative and associative properties of multiplication.	9b Using the Commutative and Associative Properties of Multiplication
9.3 Know and use the distributive property.	9c Using the Distributive Property
Topic 10 Expressions and Equations	
10.1 Record the rule for a pattern as an expression.	10a Using Patterns to Solve Word Problems
10.2 Write expressions for situations that include an unknown quantity.	10b Writing Expression
10.3 Write equations for word problems that include an unknown quantity.	10c Writing Equations
10.4 Solve simple equations for word problems that include an unknown quantity	10d Solving Equations, Example A
10.4 Solve simple equations for word problems that include an unknown quantity	10e Solving Equations, Example B

Topic 6: Use Addition or Subtraction to Compute

Topic Introduction

Complete with teacher help if needed.

1. Use mental math to add 20 + 30.

   a. How many tens are in 20? _____

   b. How many tens are in 30? _____

   c. Add the tens. _____ tens

   d. _____ tens = _____

2. What is the sum for the addition problem shown with base ten blocks?

   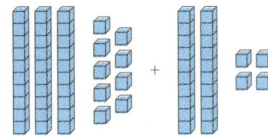

   a. Write the addition problem using numbers. _____

   b. There are 5 tens and _____ ones.

   c. Trade 10 ones for _____ ten.

   d. The sum is _____.

Objective 6.2: Use mental arithmetic to find the sum or difference of two-digit numbers.

Objective 6.4: Add 2 two-digit numbers with and without regrouping.

3. Subtract: 62 − 29.

   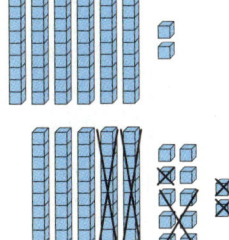

   a. Trade 1 ten for _____ ones.

   b. Cross out 2 tens and _____ ones.

   c. _____ tens _____ ones are left.

   d. The difference is _____.

4. Add: 457 + 54.

   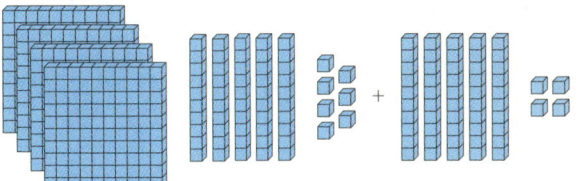

   a. How many of each type of base ten blocks?

   _____ ones _____ tens _____ hundreds

   b. Rewrite the number, trading when you can.

   _____ hundreds _____ tens _____ ones

Objective 6.5: Subtract 2 two-digit numbers with and without regrouping.

Objective 6.6: Find the sum or difference of two whole numbers up to three digits long.

Volume 3 — Level F

Lesson 6-1 — Add and Subtract Numbers

Model It

Activity 1

Use tens and ones to find 43 + 9.

Model each number. Combine the ones. Trade 10 ones for 1 ten.

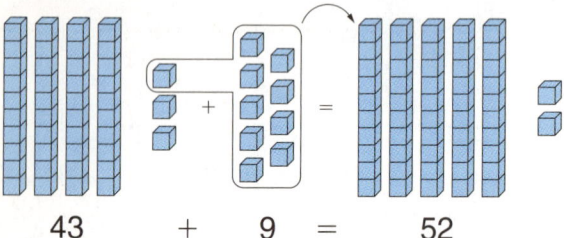

43 + 9 = 52

Practice 1

Use tens and ones to find 18 + 45.

How many ones in all? _____

After you trade 10 ones for 1 ten, there are _____ tens and _____ ones in all.

18 + 45 = _____

Activity 2

Use tens and ones to subtract 51 − 27.

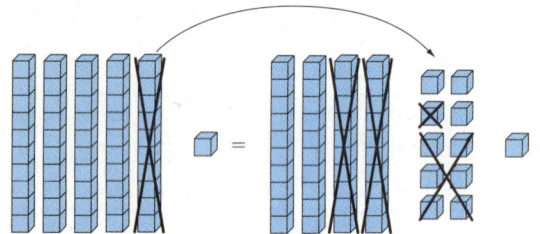

Trade 1 ten for 10 ones.
Subtract 2 tens and 7 ones.
There are 2 tens and 4 ones left.

51 − 27 = 24

Practice 2

Use tens and ones to find 60 − 8.

60 is _____ tens.

Trade 1 ten for _____ ones.

Now there are 5 tens and _____ ones.

To subtract 8, take away _____ ones.

60 − 8 = _____

On Your Own

Steve has 27 baseball cards. He gets some more for his birthday. Now he has 31 baseball cards. How many did he get for his birthday?

Steve got _____ baseball cards for his birthday.

Write About It

When you subtract, sometimes you need to trade a ten for 10 ones and sometimes you don't. When do you need to trade?

Objective 6.1: Solve addition and subtraction problems with 1- and 2-digit numbers.

Lesson 6-1 — Add and Subtract Numbers

Understand It

Example 1

Use the number line to help you subtract.
63 − 7 = ?

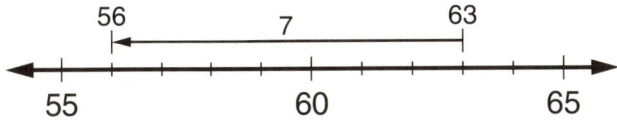

Start at 63. Move 7 units to the left. You stop at 56.
63 − 7 = 56

Practice 1

Use the number line to help you add.
59 + 6 = ?

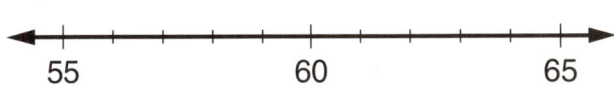

Start at _____.

Move to the right _____ units.

You stop at _____.

Example 2

Casey has 28 comic books. A friend gives him 15 more comic books. How many comic books does Casey have now?

Add to find how many in all.

Add ones. Trade ones. Add tens.

```
         1        1
  28     28      28
+ 15   + 15    + 15
        ___     ___
         3      43
```

Casey has 43 comic books.

Practice 2

Lourdes has 39 stuffed animals. She sells some at a garage sale but still has 23 left. How many did she sell?

Subtract to find how many she sold.

Lourdes sold _____ stuffed animals.

On Your Own

Brandy has 63 e-mail messages to read. She reads some but still has 27 left to read. How many e-mails did she read?

Brandy read _____ e-mails.

Write About It

Explain how you could use a number line to subtract 8 from 32.

Objective 6.1: Solve addition and subtraction problems with 1- and 2-digit numbers.

Volume 3 — 3 — Level F

Lesson 6-1 | **Add and Subtract Numbers**

Try It

1. Write the addition problem for the picture.

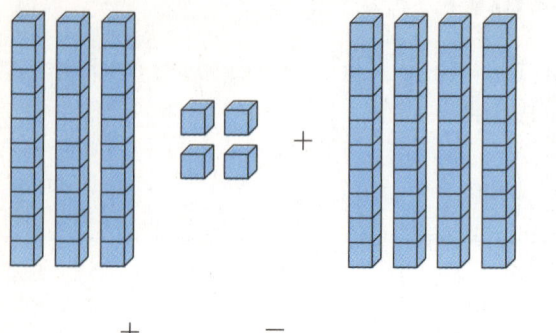

_____ + _____ = _____

2. Use the number line to find the sum or difference.

   a. 56 + 3 = _____

   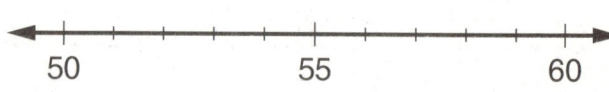

   b. 41 − 6 = _____

   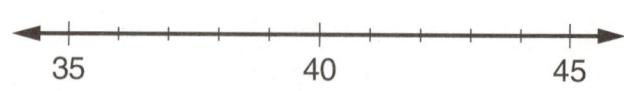

3. Add or subtract.

   a. 71
      + 4
      ____

   b. 63
      − 7
      ____

   c. 85
      +14
      ____

   d. 54
      −39
      ____

4. An alarm will ring in 45 minutes. After 19 minutes, how many minutes are left before the alarm rings? Circle the letter of the correct answer.

   A 16 minutes   B 24 minutes

   C 26 minutes   D 74 minutes

5. Alice says 73 + 18 equals 81. What mistake did she make?

6. To find 74 − 37, is it necessary to regroup? Explain.

7. Mark has some books. He gives 6 to a friend. He has 49 left. How many did he have to start with?

8. Juan buys 36 granola bars. He gives some to a friend. He has fewer than 25 left. How many bars do you think he gave to his friend?

Objective 6.1: Solve addition and subtraction problems with 1- and 2-digit numbers.

Volume 3 — 4 — Level F

Lesson 6-2: Add or Subtract Mentally

A Model It

Activity 1

Find the number that is ten more than 56.

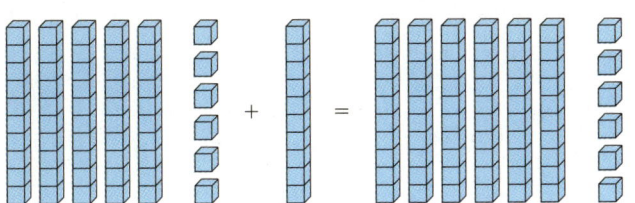

To find 10 more, add 1 more ten.
56 + 10 = 66
Ten more than 56 is 66.

Practice 1

Find the number that is twenty less than 35.

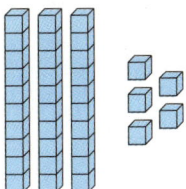

To find twenty less, subtract _____ tens.

Twenty less than 35 is _____.

Activity 2

Find the sum 42 + 48 mentally.
Break apart and make 10.

$$42 + 48$$
Break apart: 42 + 40 + 8
Make 10: 42 + 8 + 40
          50   + 40 = 90

The sum of 42 + 48 is 90.

Practice 2

Find the sum 16 + 64 mentally.
Break apart and make 10.

16 + 64
Break apart: 16 + 60 + ____
Make 10: 16 + 4 + ____
         ____ + ____ = ____

16 + 64 = _____

On Your Own

Find the sum 64 + 36 mentally.

64 + 36 = _____

Write About It

Explain how to find the sum 29 + 41 using mental math.

Objective 6.2: Use mental math to find the sum or difference of two 2-digit numbers.

Lesson 6-2 | **Add or Subtract Mentally**

B Understand It

Example 1

Find the difference 70 − 45 mentally.

Break apart each number to subtract mentally.

Break apart: 70 = 60 + 10
              45 = 40 + 5

Subtract: 60 − 40 = 20
           10 − 5 = 5
           20 + 5 = 25

The difference of 70 − 45 is 25.

Practice 1

Find the difference 80 − 25 mentally.

80 = _____ + 10

25 = _____ + _____

The difference 80 − 25 is _____.

Example 2

Subtract 83 − 62 mentally.

Break apart to subtract mentally.

Break apart: 62 = 60 + 2

Subtract: 83 − 60 = 23
           23 − 2 = 21
           83 − 62 = 21

Practice 2

Subtract 58 − 24 mentally.
Break apart to subtract mentally.

Break apart: 24 = 20 + _____

Subtract: 58 − 20 = _____

           38 − 4 = _____

           58 − 24 = _____

On Your Own

Subtract 48 − 16 mentally.

48 − 16 = _____

Write About It

Could you break apart the second number to find 56 − 39 mentally? Explain.

Objective 6.2: Use mental math to find the sum or difference of two 2-digit numbers.

Lesson 6-2 **Add or Subtract Mentally**

Try It

1. Find the number that has the following properties.

   a. ten more than 65 _____

   b. ten less than 47 _____

   c. twenty more than 58 _____

   d. thirty less than 86 _____

2. Mentally, find the sum.

   a. 58 + 22 = _____

   b. 42 + 45 = _____

   c. 24 + 75 = _____

   d. 45 + 25 = _____

3. Mentally, find the difference.

   a. 65 − 24 = _____

   b. 70 − 45 = _____

   c. 87 − 55 = _____

   d. 73 − 23 = _____

4. Use mental math to compute. Which is less than 56? Circle the letter of the correct answer.

   A 32 + 26   B 27 + 40

   C 83 − 50   D 94 − 20

5. Chin biked 35 miles one day and 15 miles the next. How far did he bike in all? Circle the letter of the correct answer.

   A 30 miles   B 35 miles

   C 40 miles   D 50 miles

6. Shamika says that it is easy to mentally add two numbers that have 5s in the ones place. Why do you think she says this?

7. To add 37 and 27 mentally Kyle used 7 + 7 = 14, 30 + 20 = 50, 50 + 4 = 54. Did he get the right answer? Explain.

8. Jane and Lonnie are playing a game. Each player makes two 2-digit numbers using each of the digits 1, 2, 3, and 4 once. The goal is to create two numbers with the greatest possible sum. Jane makes 32 and 14. Lonnie makes 24 and 31. Who wins the game? Explain.

Objective 6.2: Use mental math to find the sum or difference of two 2-digit numbers.

Volume 3   7   Level F

Lesson 6-3   Add or Subtract with Estimation

Model It

Words to Know   An **estimate** is a number close to the actual answer.

Activity 1

Use the number line to round each number. Then find the sum.

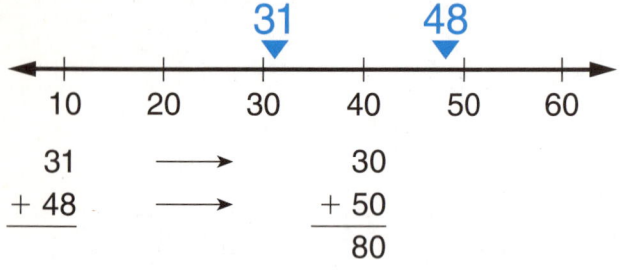

```
 31         30
+48   →   +50
          ───
           80
```

Practice 1

Use the number line to round each number. Then find the sum.

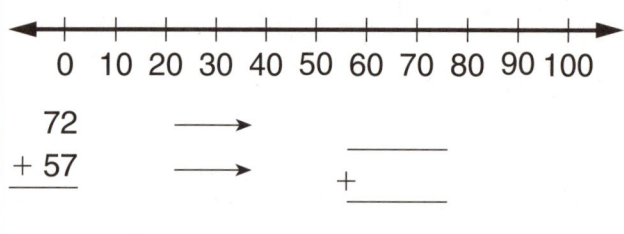

```
 72         ___
+57   →    +___
           ───
            ___
```

Activity 2

Round to estimate the difference.

```
 45         50
-16   →   -20
          ───
           30
```

Practice 2

Round to estimate the difference.

```
 68         ___
-34   →    -___
           ───
            ___
```

On Your Own

Round to estimate.

a. 46 + 19

    Estimate: _____

b. 58 − 19

    Estimate: _____

Write About It

Explain how to estimate 78 + 11 by rounding.

Objective 6.3: Estimate sums and differences of 2-digit numbers.

Volume 3   8   Level F

Lesson 6-3: Add or Subtract with Estimation

Understand It — B

Example 1

Use front-end estimation to estimate the sum.

Keep the first digit the same, and change all the digits after it to 0.

```
  43   →    40
+ 51   →  + 50
           ----
            90
```

Practice 1

Use front-end estimation to estimate the sum.

```
  56    Change the ones digit to 0.    ____
+ 37    Change the ones digit to 0.   +____
                                       ----
```

Example 2

Use front-end estimation to estimate the difference.

Keep the first digit the same, and change all the digits after it to 0.

```
  72   →    70
- 38   →  - 30
           ----
            40
```

Practice 2

Use front-end estimation to estimate the difference.

```
  68    Change the ones digit to 0.    ____
- 34    Change the ones digit to 0.   -____
                                       ----
```

On Your Own

Use front-end estimation to estimate.

62 + 34

Estimate: _____

66 − 21

Estimate: _____

Write About It

Explain how to estimate 27 + 51 by front-end estimation.

Objective 6.3: Estimate sums and differences of 2-digit numbers.

Volume 3 — Level F

Lesson 6-3 — **Add or Subtract with Estimation**

Try It

1. Round to estimate.

   a. 44 + 23   Estimate: _____

   b. 58 − 12   Estimate: _____

   c. 39 + 26   Estimate: _____

2. Use front-end estimation to estimate.

   a. 44 + 23   Estimate: _____

   b. 58 − 12   Estimate: _____

   c. 39 + 26   Estimate: _____

3. Round to estimate.

   a. 42 + 48   Estimate: _____

   b. 77 − 22   Estimate: _____

4. Use front-end estimation to estimate.

   a. 52 + 19   Estimate: _____

   b. 38 − 12   Estimate: _____

5. What is the estimate for 55 + 21 when the addends are rounded? Circle the letter of the correct answer.

   A 70   B 80

   C 90   D 100

6. A pair of earrings costs $29 and a matching necklace costs $58. About how much do the earrings and necklace cost together? Use front-end estimation. Show your work.

   _____

   _____

   The earrings and necklace cost about

   _____.

7. If each sum is estimated by rounding and front-end estimation, which sum will have the same estimate from both methods? Circle the letter of the correct answer.

   A 16 + 26   B 14 + 37

   C 21 + 73   D 48 + 35

8. Marcia was trying to figure out whether she had enough money for the jewelry in Exercise 6 or for another outfit in a store. She should **not** use front-end estimation. Why?

Objective 6.3: Estimate sums and differences of 2-digit numbers.

Lesson 6-4 — Add 2-Digit Numbers

Model It

Words to Know — **Regroup** means to exchange amounts of equal value to rename a number.

Activity 1

Use base ten blocks to find 37 + 24.

There are 11 ones in the total.

> Regroup as 1 ten and 1 one.

37 + 24 = 61

Practice 1

Use base ten blocks to find 25 + 48.

There are _____ one(s).

25 + 48 = _____

> Regroup as _____ ten(s) and _____ one(s).

Activity 2

Use base ten blocks to find 78 + 37.

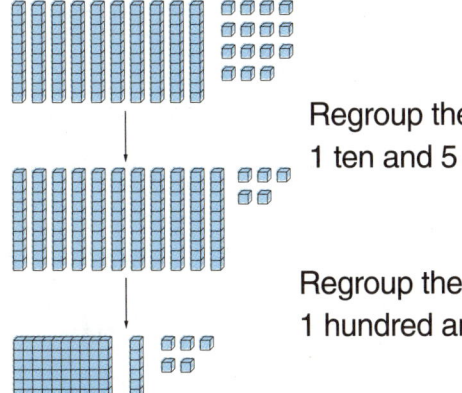

Regroup the ones as 1 ten and 5 ones.

Regroup the tens as 1 hundred and 1 ten.

78 + 37 = 115

Practice 2

Use base ten blocks to find 64 + 59.

First there are _____ ones. Regroup as _____ ten(s) and _____ ones. That makes _____ tens. Regroup as _____ hundred(s) and _____ tens.

64 + 59 = 123

On Your Own

Regroup these base ten blocks.

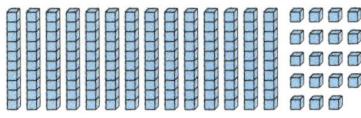

Write About It

Find the sum 72 + 39. List the base ten blocks that represent the answer.

Objective 6.4: Add two 2-digit numbers with and without regrouping.

Lesson 6-4: Add 2-Digit Numbers

B — Understand It

Example 1

Find the sum.

```
  1
  45
+ 38
-----
  83
```

There are 13 ones. Regroup as 1 ten and 3 ones. Write the 1 ten as a small 1 in the tens column.

Practice 1

Find the sum.

```
  65
+ 29
-----
```

There are _____ one(s).

Regroup as _____ ten(s) and _____ one(s).

Example 2

Find each sum. Circle the problem that needs regrouping.

```
  45              1 1
+ 32              97
----            + 58
  77            -----
                 155
```

17 ones → 1 ten, 7 ones
15 tens → 1 hundred, 5 tens

Practice 2

Find each sum. Circle the problem that needs regrouping in **two** places.

```
  52              46
+ 19            + 64
-----           -----
```

Describe the regrouping in that problem.

_____ ones → _____ ten, _____ ones

_____ tens → _____ hundred, _____ ten

On Your Own

Find each sum.

```
  68              89
+ 19            + 68
-----           -----
```

Write About It

Does 38 + 97 need regrouping? Explain.

Objective 6.4: Add two 2-digit numbers with and without regrouping.

Volume 3 — Level F

Lesson 6-4   **Add 2-Digit Numbers**

1. What sum do these base ten blocks represent?

   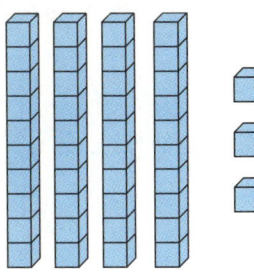

   A 17 + 23   **B** 18 + 25

   C 29 + 16   **D** 26 + 27

2. What is the missing digit?

   ```
     38
   + 45
   ----
    ? 3
   ```

   A 1   **B** 7

   C 8   **D** 9

3. Juan played for 38 minutes in the game's first half. Ben played for 46 minutes in the second half. How many minutes did they play all together? Show your work.

   They played _____ minutes.

4. Molly practiced her trumpet 37 minutes Monday and 25 minutes Tuesday. How many minutes did she practice her trumpet all together? Show your work.

   Molly practiced for _____ minutes.

5. Find each sum. Circle the problem that needs regrouping.

   ```
     42           35
   + 58         + 63
   ----         ----
   ```

6. Find each sum. Circle the problem that needs regrouping.

   ```
     25           36
   + 33         + 28
   ----         ----
   ```

7. Which of the following expressions needs regrouping?

   A 12 + 27   **B** 25 + 33

   C 49 + 10   **D** 35 + 45

8. Sam is adding 25 and 43. Does Sam need to regroup? Why or why not?

Objective 6.4: Add two 2-digit numbers with and without regrouping.

Lesson 6-5: Subtract 2-Digit Numbers

Model It

Words to Know — **Regrouping** is exchanging amounts of equal value to rename a number. You can regroup 10 ones as 1 ten. You can regroup 1 ten as 10 ones.

Activity 1

Use base ten blocks to solve 34 − 7.

Think: Can I take away 7 ones from 4 ones? No.

Regroup. Separate a ten into 10 ones.

Now there are 2 tens and 14 ones. Subtract 7 ones.

34 − 7 = 27

Practice 1

Use base ten blocks to solve 26 − 8.

Think: Can I take away 8 ones from 6 ones? _____

Regroup. Separate a ten into _____.

Now there are _____ ten and _____ ones.

Subtract 8 ones.

26 − 8 = 17

Activity 2

Solve 86 − 59.

You can't subtract 9 ones from 6 ones.

Regroup to make 7 tens and 16 ones.

Subtract 9 ones.
Subtract 5 tens.

86 − 59 = 27

Practice 2

Solve 95 − 87.

Model 95.

You can't subtract 7 ones from 5 ones.

Regroup to make _____ tens and _____ ones.

Subtract _____ ones.

Subtract _____ tens.

95 − 87 = _____

On Your Own

Find each difference. Use models to help.

```
  54        72
 − 8       − 9
 ___       ___
```

Write About It

You have 6 tens and 5 ones. You want to subtract 6 ones. Explain how to regroup.

Objective 6.5: Subtract two 2-digit numbers with and without regrouping.

Lesson 6-5: Subtract 2-Digit Numbers

B Understand It

Example 1

Subtract.

$$\begin{array}{r} \overset{6}{\cancel{7}}\overset{12}{\cancel{2}} \\ -8 \\ \hline 64 \end{array}$$

Regroup 1 ten as 10 ones. Cross out the 7 and mark 6 above it to show how many tens there are now.
Then cross out the 2 and mark 12 above it to show how many ones there are now.
Subtract ones. Subtract tens.

Practice 1

Subtract.

$$\begin{array}{r} 84 \\ -7 \\ \hline \end{array}$$

Regroup. Now there are
_____ tens and
_____ ones.

Subtract ones.
Subtract tens.

Example 2

Circle the problem that requires regrouping. Find the difference.

$$\begin{array}{r} 54 \\ -24 \\ \hline \end{array}$$

$$\begin{array}{r} \overset{4}{\cancel{5}}\overset{14}{\cancel{4}} \\ -28 \\ \hline 26 \end{array}$$ (circled)

Regroup. Now there are 4 tens and 14 ones.
Subtract 8 ones.
Subtract 2 tens.

Practice 2

Circle the problem that requires regrouping. Find the difference.

$$\begin{array}{r} 75 \\ -49 \\ \hline \end{array}$$

Regroup. Now there are _____ tens and _____ ones.

Subtract _____ ones.

Subtract _____ tens.

$$\begin{array}{r} 75 \\ -44 \\ \hline \end{array}$$

On Your Own

Find each difference. Show your work.

$$\begin{array}{r} 80 \\ -46 \\ \hline \end{array} \qquad \begin{array}{r} 96 \\ -78 \\ \hline \end{array} \qquad \begin{array}{r} 82 \\ -71 \\ \hline \end{array}$$

Write About It

Will you regroup to solve this problem? Explain how you know.

$$\begin{array}{r} 95 \\ -27 \\ \hline \end{array}$$

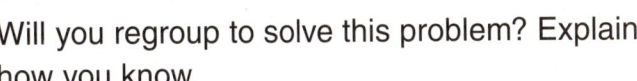

Objective 6.5: Subtract two 2-digit numbers with and without regrouping.

Lesson 6-5

Subtract 2-Digit Numbers

C Try It

1. Find each difference. Show your work.

a. $\begin{array}{r} 51 \\ -\ 4 \\ \hline \\ \end{array}$ ____

b. $\begin{array}{r} 50 \\ -\ 0 \\ \hline \\ \end{array}$ ____

2. Find each difference. Show your work.

a. $\begin{array}{r} 65 \\ -14 \\ \hline \end{array}$

b. $\begin{array}{r} 90 \\ -23 \\ \hline \end{array}$

3. Show how to subtract 14 from this model.

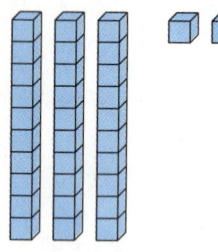

Write the problem and the answer.

____

4. Avery is regrouping, and needs to exchange a ten for ones. After regrouping, what blocks will he have? Circle the letter of the correct answer.

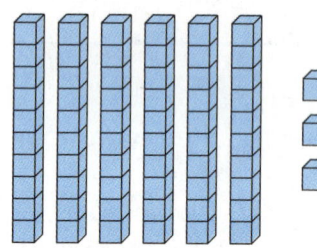

A 5 tens, 3 ones **B** 6 tens, 3 ones

C 5 tens, 13 ones **D** 6 tens, 13 ones

5. A restaurant manager ordered 87 bottles of ketchup and 59 bottles of mustard. How many more bottles of ketchup did she order? Show your work.

She ordered ______ more ketchup bottles.

6. What is the missing number? Explain.

$\begin{array}{r} 62 \\ -26 \\ \hline ?6 \end{array}$

7. Circle the letter of the problem that requires regrouping.

A $\begin{array}{r} 61 \\ -42 \\ \hline \end{array}$

B $\begin{array}{r} 47 \\ -21 \\ \hline \end{array}$

C $\begin{array}{r} 35 \\ -23 \\ \hline \end{array}$

D $\begin{array}{r} 56 \\ -40 \\ \hline \end{array}$

8. Brent and his dad caught 48 fish. He threw back 19 and kept the rest. How many fish did they keep? Show your work.

They kept ______ fish.

Objective 6.5: Subtract two 2-digit numbers with and without regrouping.

Lesson 6-6 — **Add or Subtract 3-Digit Numbers**

Model It

Activity 1

Use base ten blocks to solve this problem.

256 + 56 = ?

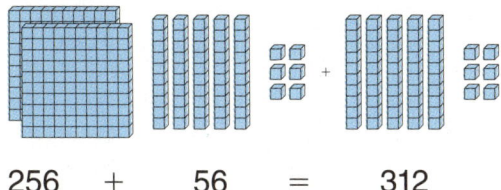

256 + 56 = 312

Combine ones, tens, and then hundreds. Trade for a bigger block when you can.

Practice 1

Use base ten blocks to solve this problem.

537 + 458 = ?

How many ones in all? _____

Trade _____ ones for _____ tens.

Now there are _____ tens and _____ ones.

There are _____ hundreds.

537 + 458 = _____

Activity 2

Use base ten blocks to help you solve this problem.

100 − 67 = ?

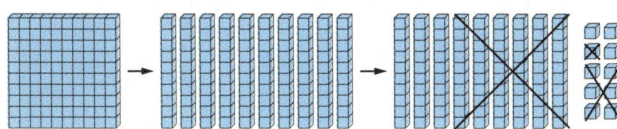

```
  H   T   O
      9
  ̷1  1̷0  10
      ̷0   ̷0
  −   6   7
      3   3
```

Practice 2

Use base ten blocks to help you solve this problem.

407 − 115 = ?

```
  H   T   O
  4   0   7
− 1   1   5
```

On Your Own

Add. Remember to write the regrouping number.

```
  H   T   O
  4   1   2
 +3   9   6
```

Write About It

What is the first step when you subtract 520 from 700 using base ten blocks?

Objective 6.6: Find the sum or difference of two whole numbers up to three digits long.

Lesson 6-6 Add or Subtract 3-Digit Numbers

Example 1

Use the number line to help you subtract.

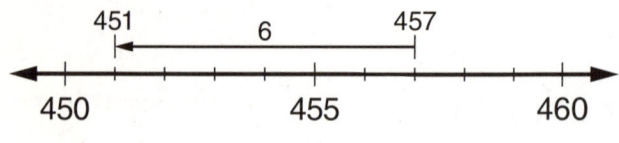

457 − 6 = ?

Start at 457. Move 6 to the left. You stop at 451.

457 − 6 = 451

Practice 1

Use the number line to help you add.

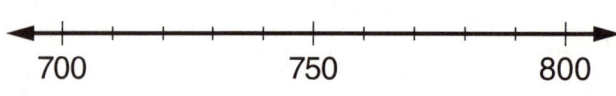

730 + 60 = ?

Start at _____. Move _____ to the right.

You stop at _____.

Example 2

Damian has 279 trading cards. He buys a pack of 35 cards. How many does he have in all?

To find how many in all, add 279 + 35.

```
           1        1 1
  279     279      279
+  35   +  35    +  35
                 ─────
           4      314
```

Damian has 314 trading cards.

Practice 2

The Friendship Club made 105 bracelets to sell during the parent open house. The members counted 67 bracelets after the event. How many did they sell?

Subtract to find how many were sold.

```
 H  T  O

 1  0  5
−   6  7
─────────
```

_____ bracelets were sold.

On Your Own

Carla has $197 in her savings account. She deposits $255 in the account. How much is in her account?

Carla has $_____ in her account.

Write About It

Explain how to use a number line to subtract 300 from 700.

Objective 6.6: Find the sum or difference of two whole numbers up to three digits long.

Lesson 6-6 — Add or Subtract 3-Digit Numbers

1. Write the subtraction problem for the picture.

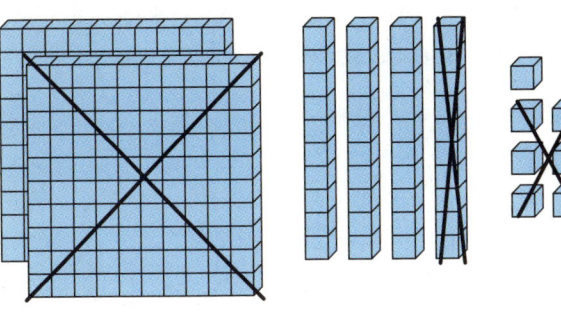

___ − ___ = ___

2. Use the number line to find the sum or difference.

   a. 232 + 7 = ___

   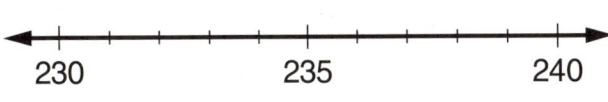

   b. 800 − 40 = ___

   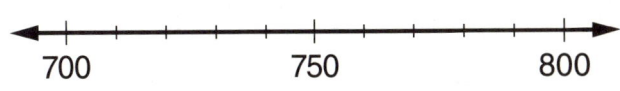

3. Add or subtract.

   a. 273
      + 88

   b. 580
      − 48

   c. 594
      + 316

   d. 660
      − 326

4. In which subtraction would you need to trade 1 hundred for 10 tens? Circle the letter of the correct answer.

   A 983 − 261   B 638 − 265

   C 792 − 563   D 385 − 128

5. Kyle says 543 + 269 equals 702. Is his answer correct? Explain why or why not.

6. Is 154 + 32 the same as 100 + 50 + 30 + 4 + 2? Why?

7. In which addition would you need to regroup 10 ones as 1 ten?

   A 433 + 215   B 182 + 364

   C 508 + 330   D 627 + 234

8. Mr. Washington walked 83 blocks on Saturday. Over the two-day weekend he walked 310 blocks. How far did he walk on Sunday?

Objective 6.6: Find the sum or difference of two whole numbers up to three digits long.

Lesson 6-7: Multidigit Numbers

Model It

Activity 1

Find 516 + 196. Explain each step.
You can use base ten blocks to help.

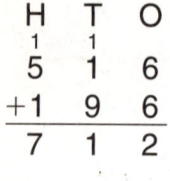

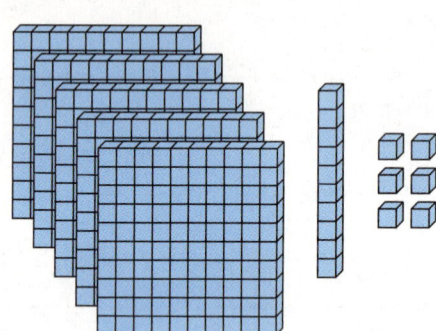

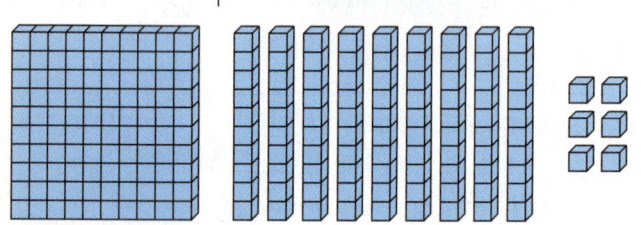

Practice 1

Add 649 + 688. Explain each step.
You can use base ten blocks to help.

```
  H  T  O
     
  6  4  9
 +6  8  8
```

Add ones. There are _____ ones. Regroup.

Add tens. There are _____ tens. Regroup.

Add hundreds. There are _____ hundreds. Regroup.

Activity 2

Estimate 5,624 + 72,738. Then find the exact sum.
Round each number to the thousands place.

5,624 → 6,000   72,738 → 73,000

Write the problem vertically.
Line up the places.
Add in each place. Regroup when necessary.

```
TTh Th  H  T  O           TTh Th  H  T  O
                                 1     1
     6, 0  0  0                5, 6  2  4
   +7 3, 0  0  0             +7 2, 7  3  8
    7 9, 0  0  0              7 8, 3  6  2
```

Practice 2

Solve 202,387 + 82,796.

```
  202,387
+  82,796
```

Estimate to check.

```
  200,000
+  80,000
```

On Your Own

Add 128 + 4,823. Estimate to check.

_____

Write About It

When do you need to regroup?

Objective 6.7: Use standard algorithms to add and subtract multidigit numbers.

Volume 3   Level F

Lesson 6-7 | Multidigit Numbers

B Understand It

Example 1

Subtract: 345 − 167. Explain your work.

```
  H   T   O
      13
  2   3   15
  3̸   4̸   5̸
− 1   6   7
  1   7   8
```

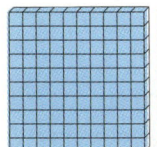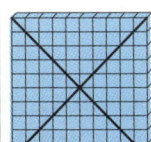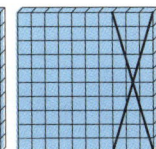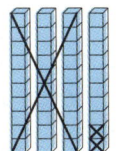

Practice 1

Subtract: 713 − 346.
Use base ten blocks if needed.

```
  H   T   O

  7   1   3
− 3   4   6
```

Example 2

Find 924,713 − 62,962. Estimate to check.

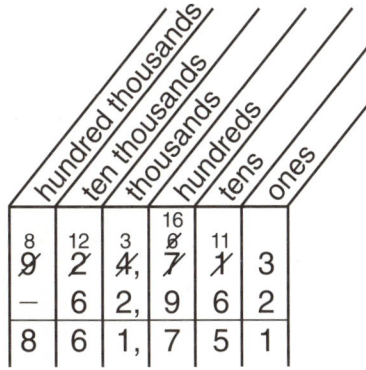

Practice 2

Find 863,378 − 28,728. Estimate to check.

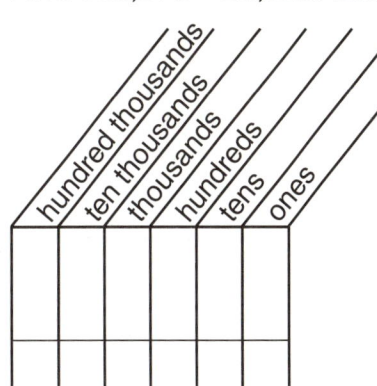

On Your Own

Subtract: 23,635 − 10,473.

_____

Estimate to check.

_____

Write About It

Give a real-world example to show the idea of trading in subtraction.

Objective 6.7: Use standard algorithms to add and subtract multidigit numbers.

Lesson 6-7 **Multidigit Numbers**

1. Subtract 1526 − 230. Estimate to check. Show your work.

2. Solve.

   a.   5,386
      + 6,450

   b.   1,017
      + 2,619

   c.   33,027
      − 21,745

   d.   34,181
      − 24,481

3. Subtract. Estimate to check.

   a.   31,222
      − 27,580

   b.   702,802
      −   2,053

4. One year Mrs. Rodriquez earned $37,632. She paid $5,629 in taxes. What was her income after taxes?

   A $32,017   B $33,144

   C $32,003   D $32,013

5. What is 12,500 less than 73,956?

   A 93,556   B 71,556

   C 68,556   D 61,456

6. The area of California is 163,696 square miles. The area of Oregon is 98,386 square miles. What is the difference of their areas?

   _____

7. A marathon swimmer swam the lengths of the Yangtze River (4,003 km), the Mississippi River (3,797 km), and the Danube River (3,004 km). What is his total swimming distance in these three rivers?

   _____
   _____

8. When adding two numbers, do you ever trade 30 ones for 3 tens? Explain why or why not.

Objective 6.7: Use standard algorithms to add and subtract multidigit numbers.

Topic 6: Use Addition or Subtraction to Compute

Topic Summary

Choose the correct answer. Explain how you decided.

1. Stephanie has a coin collection. If she collects 17 more coins, she will have a total of 515 coins. How many coins are in her collection now?

    A 488

    B 498

    C 512

    D 532

    _____

    _____

    _____

2. In 1860, the population of California was 379,994. Ten years earlier, in 1850, the population was 92,597. How many more people lived in California in 1860 than in 1850?

    A 267,377

    B 287,397

    C 287,497

    D 472,591

    _____

    _____

    _____

Objective: Review addition and subtraction.

Topic 6: Use Addition or Subtraction to Compute

Mixed Review

1. Find the sum.

a. 38 + 13 = _____

b. 16 + 44 = _____

c. 81 + 19 = _____

Volume 3, Lesson 6-1

2. Write the number six thousand, two hundred eight in standard form.

_____

Volume 1, Lesson 2-1

3. Use the cubes to model 16 − 4.

16 − 4 = _____

Volume 2, Lesson 3-3

4. Which number is a multiple of 7? Circle the letter of the correct answer.

A 17   B 21

C 29   D 36

Volume 1, Lesson 4-3

5. List the factors of 18.

_____

Volume 2, Lesson 4-5

6. Explain how to add 26 + 34 mentally.

Volume 3, Lesson 6-2

7. Explain why 42 is **not** divisible by 5.

Volume 2, Lesson 5-3

8. There are 11 players on a soccer team. Aaron needs 5 more players on his team. How many players are on his team now?

Volume 2, Lesson 3-1

Objective: Maintain concepts and skills.

Topic 7: Use Multiplication to Compute

Topic Introduction

Complete with teacher help if needed.

1. Solve.

   a. 4 × 3 = _____

   b. Complete the pattern.

   40 × 3 = _____

   400 × 3 = _____

   4,000 × 3 = _____

2. Use a number line to estimate 5 × 37.

   ```
   ←––|––|––|––|––|––→
      0  10 20 30 40 50
   ```

   a. 5 × 37 → 5 × _____

   b. So, 5 × 37 is about _____.

Objective 7.1: Solve problems involving multiplying multiples of 10.

Objective 7.2: Estimate products by rounding factors and using mental math techniques.

3. Multiply 13 × 7.

   a. Break 13 into tens and ones.

   _____

   b. _____ × 7 = _____

   c. _____ × 7 = _____

   d. Add. _____ + _____ = _____

4. Find the product.

   a. 70 × 8 = _____

   b. 6 × 29 = _____

   c. 16 × 8 = _____

   d. 300 × 5 = _____

Objective 7.3: Multiply multidigit numbers by 1-digit numbers.

Objective 7.4: Solve problems choosing between mental math, paper and pencil, or estimation.

Volume 3 — Level F

Lesson 7-1: Multiply Multiples of 10

Model It — A

Words to Know — A **multiple** is the product of a given number and another whole number.

Activity 1

2 × 7 = 1 4
2 × 7 0 = 1 4 0
2 × 7 0 0 = 1 4 0 0
2 × 7 0 0 0 = 1 4 0 0 0

Practice 1

9 × 4 = _____
9 × 4 0 = _____
9 × 4 0 0 = _____
9 × 4 0 0 0 = _____

Activity 2

5,000 × 6 = ?

5 × 6 = 30
50 × 6 = 300
500 × 6 = 3,000
5,000 × 6 = 30,000

The product of 5,000 × 6 is 30,000.

Practice 2

5 × 4,000 = ?

5 × 4 = _____
5 × 40 = _____
5 × 400 = _____
5 × 4,000 = _____

The product of 5 × 4,000 is _____.

On Your Own

a. 8 × 5 = _____

8 × 50 = _____

8 × 500 = _____

8 × 5,000 = _____

b. 5 × 7 = _____

5 × 70 = _____

5 × 700 = _____

5 × 7,000 = _____

Write About It

What pattern can you use to solve any of these problems: 60 × 30, 6 × 300, 600 × 3?

Objective 7.1: Solve problems involving multiplying multiples of 10.

Volume 3 — 26 — Level F

Lesson 7-1: Multiply Multiples of 10

Understand It (B)

Example 1

$60 \times 300 = ?$

Basic fact: $6 \times 3 = 18$

6×10 3×100 $18 \times 10 \times 100$

$60 \times 300 = 18,000$

6 is multiplied by 10 to make 60.

3 is multiplied by 100 to make 300.

So multiply the product by 10 and by 100, or by 1,000.

The product of 60×300 is 18,000.

Practice 1

$400 \times 40 = ?$

Basic fact: $4 \times 4 =$ _____

4 is multiplied by _____ to make 400.

4 is multiplied by _____ to make 40.

So multiply the product by _____ and by _____, or by _____.

$400 \times 40 =$ _____

Example 2

$500 \times 200 = ?$

Basic fact: $5 \times 2 = 10$

Multiply by 100 and by 100, or by 10,000.

$500 \times 200 = 100,000$

Practice 2

$500 \times 800 = ?$

Basic fact: _____ $\times$ _____ = _____

Multiply by _____ and by _____, or by 10,000.

$500 \times 800 =$ _____

On Your Own

a. $90 \times 300 =$ _____

b. $8,000 \times 80 =$ _____

c. $600 \times 500 =$ _____

Write About It

Write a word problem that you can solve by starting with 5×3 and multiplying that product by 100 and by 10.

Objective 7.1: Solve problems involving multiplying multiples of 10.

Lesson 7-1 — **Multiply Multiples of 10**

Try It

1. [7] × [6] = _____

   [7] × [6][0] = _____

   [7] × [6][0][0] = _____

   [7] × [6][0][0][0] = _____

2. 5 × 2,000 = ?

   5 × 2 = _____

   5 × 20 = _____

   5 × 200 = _____

   5 × 2,000 = _____

   The product of 5 × 2,000 is _____.

3. 90 × 800 = ?

   Basic fact: 9 × 8 = _____

   Multiply the product by _____.

   90 × 800 = _____

4. Solve.

   a. 6,000 × 6 = _____

   b. 800 × 60 = _____

   c. 700 × 300 = _____

5. Circle the letter of the correct product.
   400 × 50 = ?

   A 200
   B 2,000
   C 20,000
   D 200,000

6. Which product will have 4 zeros? Circle the letter of the correct answer.

   A 30 × 300
   B 500 × 600
   C 50 × 900
   D 4,000 × 60

7. Which product will have 5 zeros? Circle the letter of the correct answer.

   A 70,000 × 3
   B 80 × 5,000
   C 500 × 2,000
   D 900 × 60

8. Mika has 4 photo albums. Each album holds 300 pictures. How many pictures are in Mika's photo albums? Explain how you got your answer.

Objective 7.1: Solve problems involving multiplying multiples of 10.

Volume 3 — 28 — Level F

Lesson 7-2: Estimate Products

Model It — A

Words to Know   **Round** means to change the digits in a number to make a similar but easier to use number. An **estimate** is a number close to the actual answer.

Activity 1

Use a number line to estimate 4 × 37.

Round 37 to the greatest place.
37 is between 3 tens and 4 tens.
It is closer to 40 than to 30.

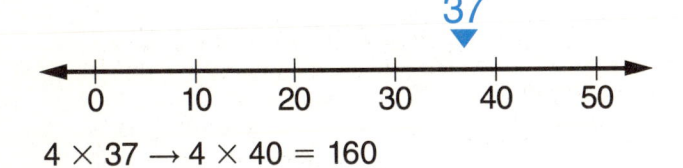

4 × 37 → 4 × 40 = 160

So 4 × 37 is about 160.

Practice 1

Use a number line to estimate 8 × 293.

Round 293 to the greatest place.
293 rounds to _____.

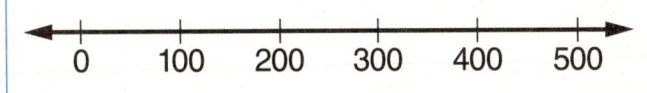

8 × 293 → 8 × _____ = _____

So 8 × 293 is about _____.

Activity 2

Estimate 368 × 23. Use number lines.

Round 368 to the greatest place.
368 rounds to 400.

Round 23 to the greatest place.
23 rounds to 20.

368 × 23 → 400 × 20 = 8,000
So, 368 × 23 is about 8,000.

Practice 2

Estimate 715 × 259. Use number lines.

715 rounds to _____.

259 rounds to _____.

715 × 259 → _____ × _____ = _____

So, 715 × 259 is about _____.

On Your Own

Use a number line to estimate.

a. 6 × 2,712

   6 × _____ = _____

b. 78 × 21

   _____ × _____ = _____

Write About It

Look at Activity 1 and Practice 1. Why do you think we didn't round the 6 or the 8?

Objective 7.2: Estimate products by rounding factors and using mental math techniques.

Volume 3   Level F

Lesson 7-2: Estimate Products

B Understand It

Example 1

Use rounding rules to estimate 371 × 5.

1. Round 371 to the nearest hundred. It will be 300 or 400.
2. Underline the hundreds digit.
3. Circle the digit to its right.
4. If the circled digit is 5 or greater, increase the underlined digit by 1.

   3⑦1   7 > 5, so change 3 to 4.
371 × 5 → 400 × 5 = 2,000
So 371 × 5 is about 2,000.

Practice 1

Use rounding rules to estimate 1,396 × 8.

1. Round 1,396 to the nearest _____. It will be _____ or _____.
2. Underline _____.
3. Circle _____.
4. Is the circled digit 5 or greater? _____
   1,396 rounds to _____.

1,396 × 8 → _____ × 8 = _____

So 1,396 × 8 is about _____.

Example 2

Use rounding to estimate 65 × 421.

Round 65 to the nearest ten.
65 rounds up to 70.

Round 421 to the nearest hundred.
421 rounds down to 400.

Multiply: 70 × 400 = 28,000

So 65 × 421 is about 28,000.

Practice 2

Use rounding to estimate 53 × 584.

Round 53 to the nearest _____.
53 rounds _____ to _____.

Round 584 to the nearest _____.
584 rounds _____ to _____.

Multiply: 50 × 600 = 3,000

So 53 × 584 is about _____.

On Your Own

Estimate.

a. 7 × 6,514

   7 × _____ = _____

b. 37 × 81

   _____ × _____ = _____

Write About It

What is wrong with this equation?
6 × 5,123 = 6 × 5,000 = 30,000

Objective 7.2: Estimate products by rounding factors and using mental math techniques.

Lesson 7-2 | **Estimate Products**

Try It

1. Estimate 172 × 3. Use the number line.

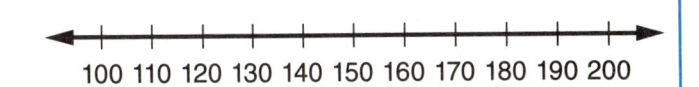

100 110 120 130 140 150 160 170 180 190 200

172 rounds to _____.

172 × 3 → _____ × 3 = _____

So 172 × 3 is about _____.

2. Estimate 64 × 85 using mental math.

64 rounds to _____.

85 rounds to _____.

64 × 85 → _____ × _____ = _____

So 64 × 85 is about _____.

3. Estimate 283 × 315 by rounding each factor to the nearest hundred.

283 × 315 → _____

4. Match each expression with its estimated product.

27 × 42	1,600
4 × 358	1,200
31 × 24	800
37 × 19	600

5. Estimate.

a. 67 × 8

_____ × 8 = _____

b. 46 × 831

_____ × _____ = _____

6. 429 × 75 is about

A 28,000   B 32,000

C 35,000   D 40,000

7. Which expression has an estimated product of 3,600? Circle the letter of the correct answer.

A 46 × 91   B 59 × 64

C 921 × 43   D 6,241 × 58

8. Write the steps to estimate 514 × 782.

Objective 7.2: Estimate products by rounding factors and using mental math techniques.

Lesson 7-3: Multiply 1-Digit Numbers

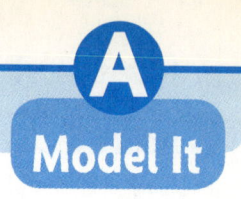

A Model It

Activity 1

Use base ten blocks to find 314 × 2.

There are 6 hundreds.
There are 2 tens.
There are 8 ones.
314 × 2 = 628

Practice 1

Use base ten blocks to find 223 × 3.

How many hundreds? _____

How many tens? _____

How many ones? _____

223 × 3 = _____

Activity 2

Use base ten blocks to find 2 × 147.

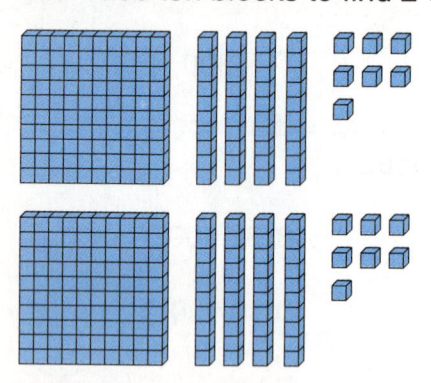

There are 2 hundreds.
There are 8 tens.
There are 14 ones.
Regroup the ones 1 more ten and 4 ones.
2 × 147 = 294

Practice 2

Use base ten blocks to find 124 × 4.

How many hundreds? _____

How many tens? _____

How many ones? _____

Regroup the ones as _____ more ten and _____ ones.

124 × 4 = _____

On Your Own

Draw a picture to show 132. Then multiply 132 × 2.

132 × 2 = _____

Write About It

Explain how to find 2,143 × 4.

Objective 7.3: Multiply multidigit numbers by one-digit numbers.

Volume 3 — 32 — Level F

Lesson 7-3 — Multiply 1-Digit Numbers

B Understand It

Example 1

Multiply 1,524 × 3.
Use place value.

1,524 = 1,000 + 500 + 20 + 4
3 × 1,000 = 3,000
3 ×   500 = 1,500
3 ×   20 =   60
3 ×   4 =   12

              4,572

1,524 × 3 = 4,572

Practice 1

Multiply 2,253 × 4.
Use place value.

2,253 = 2,000 + _____ + _____ + _____
4 × 2,000 = 8,000

4 × _____ = _____

4 × _____ = _____

4 × _____ = _____

2,253 × 4 = _____

Example 2

Carla rode her bicycle 17 miles a day for a week. How far did she ride in that week?

There are 7 days in a week.

Multiply 7 by each digit in 17.

```
   4
  17
×  7
----
 119
```

She rode her bicycle 119 miles that week.

Practice 2

A theater holds 382 people. A concert was sold out every day for a week. How many people saw the concert?

_____ people saw the concert.

On Your Own

Which of the following two products is greater than 1,000: 337 × 4 or 498 × 2? Solve.

Write About It

Explain the regrouping you do when you multiply 2,236 × 4.

Objective 7.3: Multiply multidigit numbers by one-digit numbers.

Volume 3 — 33 — Level F

Lesson 7-3   Multiply 1-Digit Numbers

C Try It

1. Write the multiplication problem. Solve.

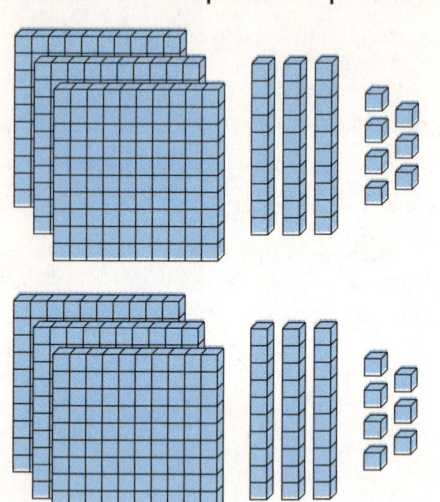

_____

2. Multiply.

a. 42 × 3 = _____

b. 3 × 322 = _____

c. 1,324 × 2 = _____

d. 2,312 × 3 = _____

3. Multiply. Regroup as needed.

a. 76 × 8 = _____

b. 4 × 843 = _____

c. 3,451 × 7 = _____

d. 5,264 × 6 = _____

4. What is 364 × 5? Circle the letter of the correct answer.

A 1,820   B 65

C 1,500   D 1,532

5. Correct each mistake.

a. 438 × 4 = 1,622

```
  438
×   4
```

b. 438 × 4 = 752

```
  1 3
  438
×   4
  752
```

6. Kim rode her bike 8 miles each day in August. She decided she had biked 2,408 miles. Is her answer reasonable?

7. Correct the mistake.

4 × 2,368 → 4 × 2

+ 4 × 3

+ 4 × 6

+ 4 × 8

8. Explain two different ways to find the product 213 × 3.

Objective 7.3: Multiply multidigit numbers by one-digit numbers.

Volume 3   Level F

Lesson 7-4 | **Choose a Method**

A Model It

Words to Know Mental math means solving math problems in your head.

Activity 1

6 × 42 = ?
Since you know basic facts with 6, use place value. Break apart and multiply.

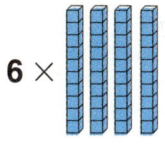

6 × [rod] [unit] → 6 × [rod] plus 6 × [unit]

6 × 40 = 240   6 × 2 = 12
240 + 12 = 252

6 × 42 = 252

Practice 1

8 × 53 = ?
Since you know basic facts with 8, use place value. Break apart and multiply.

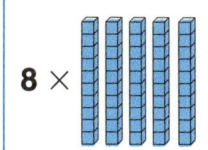

8 × [rod] [unit] → 8 × _____ plus 8 × _____

8 × 50 = 400   8 × 3 = 24
400 + 24 = 424

8 × 53 = 424

Activity 2

The freshman class is collecting food for the food bank. There are 227 freshmen. If each freshman brings 25 cans, how many cans will they donate?

Since the numbers are large, **estimate first**:
227 × 25 → 200 × 30 = 6,000

Then **multiply**:
```
   2 2 8
  ×  2 5
  1 1 3 5
+ 4 5 4 0
  5, 6 7 5
```
The answer is close to the estimate.

They will donate 5,675 cans.

Practice 2

There are 36 rows of seats in the auditorium. 121 seats are in each row. How many seats are in the auditorium?

_____ × _____
  ↓   ↓

Estimate: _____ × _____ = _____

Multiply: _____

Is the answer close to the estimate? _____

There are _____ seats.

On Your Own

Estimate. Then multiply.

517 × 94 → _____ × _____ = _____

517 × 94 = _____

Is the answer close to the estimate? _____

Write About It

Explain what it means to "break apart and multiply."

Objective 7.4: Solve problems by choosing between mental math, paper and pencil, or estimation.

Lesson 7-4 — Choose a Method

Understand It

Example 1

One ticket to the aquarium costs $23. What is the cost for a group of 7 people?

Write the problem. $23 × 7

Break apart, or estimate and solve? There is a one-digit factor, so you can break apart and multiply mentally.

20 × 7 plus 3 × 7 = 140 + 21 = 161

The cost is $161.

Practice 1

The Sullivans drove for 4 hours at 55 miles per hour. How far did they drive?

Write the problem. _____

Break apart, or estimate and solve?

_____

Show your work.

The Sullivans drove _____ miles.

Example 2

Violet works at the ice cream shop. She earns $8 an hour. She works 28 hours each week. About how much does she earn each week?

Estimate, since the problem asks *about* how much she earns.

$8 × 28 → $8 × 30 = $240

Think: 8 × 3 = 24, so 8 × 30 = 240.

Violet earns about $240 a week.

Practice 2

Jill jogs for 33 minutes each day. Approximately how many minutes will she jog in 2 months?

What word tells you to estimate?

_____

Show your work.

Jill will jog about _____ minutes.

On Your Own

T-shirts cost $18.50. There are 72 students in the band. About how much will it cost to buy each band member a T-shirt?

Choose a method.

break apart   estimate   estimate and solve

It will cost about $ _____.

Write About It

How do you decide which method to use to solve a problem?

Objective 7.4: Solve problems by choosing between mental math, paper and pencil, or estimation.

Lesson 7-4: Choose a Method

Try It

1. 683 × 71 = ?

   Estimate. Then multiply.

   683   ×   71
   ↓   ↓

   _____ × _____ = _____

   The product of 683 × 71 is _____.

2. The band sold tickets for $8 each. They sold 518 tickets. About how much did the band make in ticket sales?

   The band made about $ _____.

3. Tommy read a book in 6 hours. He read at a rate of 24 pages an hour. How many total pages did he read?

   Choose a method.

   break apart   estimate   estimate and solve

   Explain your choice.

   Tommy read _____ pages.

4. The crayon factory produces 7,594 crayons a minute. About how many crayons is this in an hour? (1 hour = 60 minutes)

   Choose a method.

   break apart   estimate   estimate and solve

   Explain your choice.

   The factory produces about _____ crayons in an hour.

5. The deli ordered 7 boxes of paper cups. Each box holds 324 cups. About how many cups did the deli order? Circle the letter of the correct answer.

   A 2,268   B 2,450

   C 2,100   D 2,800

6. Which problem is easier to solve mentally: 106 × 9 or 4 × 27? Explain your choice.

7. Write a multiplication problem that can be solved by breaking apart and multiplying. Show the solution.

8. Write a multiplication problem that can be solved using mental math. Show the solution.

Objective 7.4: Solve problems by choosing between mental math, paper and pencil, or estimation.

Topic 7: Use Multiplication to Compute

Topic Summary

Choose the letter of the correct answer. Explain how you decided.

1. There are 48 boxes to ship. If each box holds 7 jars, about how many jars will be shipped?

   A 336 jars

   B 350 jars

   C 280 jars

   D 400 jars

   _____
   _____
   _____

2. A scrapbook has 65 pages. Each page holds 6 pictures. How many pictures are in the scrapbook?

   A 71

   B 360

   C 390

   D 420

   _____
   _____
   _____

Objective: Review skills related to multiplication.

Topic 7: Use Multiplication to Compute

Mixed Review

1. Complete the pattern.

a. 6 × 7 = _____

b. 60 × 7 = _____

c. 60 × 70 = _____

Volume 3, Lesson 7-1

2. Write the place of the underlined digit in 6,3<u>5</u>1.

_____

Volume 1, Lesson 2-2

3. Nathan added 69 + 25 and got 84. Is Nathan correct? Why or why not?

_____

_____

_____

Volume 3, Lesson 6-4

4. Jerry has 56 football cards. He gave some to his brother. Jerry now has 37 cards. Which number sentence shows how many cards Jerry gave his brother? Circle the letter of the correct answer.

A 56 + 37 = 93

B 37 + 56 = 93

C 37 − 56 = 19

D 56 − 37 = 19

Volume 2, Lesson 3-1

5. What is six million, eight hundred thirty-seven thousand, one hundred five in standard form?

A 6,873,150

B 6,837,150

C 6,837,105

D 6,807,105

Volume 1, Lesson 2-3

6. Find each product.

a. 9 × 8 = _____

b. 7 × 7 = _____

c. 9 × 7 = _____

d. 8 × 6 = _____

Volume 2, Lesson 4-3

7. Round to the nearest hundred.

a. 818 _____

b. 4,692 _____

c. 5,319 _____

Volume 1, Lesson 2-4

8. List the factors of 24.

_____

Volume 2, Lesson 4-5

Objective: Maintain concepts and skills.

Topic 8: Use Division to Compute

Topic Introduction

Complete with teacher help if needed.

1. Find the quotient.

a. 42 ÷ 7 = _____

b. Complete the pattern.

420 ÷ 7 = _____

420 ÷ 70 = _____

4,200 ÷ 700 = _____

Objective 8.1: Solve problems involving dividing multiples of 10.

2. Estimate the quotient.

a. Use compatible numbers to estimate 2,318 ÷ 6.

b. Is 23 ÷ 6 a basic fact? _____

c. What number is close to 23 and a multiple of 6? _____

d. 2,318 ⟶ _____ ÷ 6 = _____

Objective 8.2: Estimate quotients by rounding factors and using mental math techniques.

3. Divide.

a. Use base ten blocks to divide 108 ÷ 9.

b. 108 ÷ 9 = _____

Objective 8.3: Divide multidigit numbers by 1-digit numbers with and without remainders.

4. A hockey coach purchased 30 hockey jerseys. If 5 jerseys are packaged in a box, how many boxes will it take to ship all of the hockey jerseys?

a. Will you multiply or divide? _____

b. 30 ÷ 5 = _____

Objective 8.5: Solve word problems by using either multiplication or division.

Lesson 8-1 Divide Multiples of 10

A Model It

Words to Know dividend → 40 ÷ 8 = 5 ← quotient
divisor ↑

Activity 1

Find the quotients.

8 ÷ 4 = 2
80 ÷ 4 = 20
800 ÷ 4 = 200
8,000 ÷ 4 = 2,000

Practice 1

Find the quotients.

9 ÷ 3 = _____
90 ÷ 3 = _____
900 ÷ 3 = _____
9,000 ÷ 3 = _____

Activity 2

Find the quotients. 80 ÷ 40 = 2, 800 ÷ 40 = 20, and 8,000 ÷ 40 = 200

Practice 2

Find the quotients: 90 ÷ 30, 900 ÷ 30, and 9,000 ÷ 30.

90 ÷ 30 = _____
900 ÷ 30 = _____
9,000 ÷ 30 = _____

On Your Own

Solve.

a. 60,000 ÷ 2,000 = _____

b. 12,000 ÷ 300 = _____

c. 4,000 ÷ 20 = _____

Write About It

These problems have the same answer. Why?
12 ÷ 4 = 3, 120 ÷ 40 = 3, and 1,200 ÷ 400 = 3

Objective 8.1: Solve division problems with multiples of 10.

Lesson 8-1: Divide Multiples of 10

B Understand It

Example 1

Find each quotient.

2,400 ÷ 8 = 300

2,400 ÷ 80 = 30

2,400 ÷ 800 = 3

Basic fact:
24 ÷ 8 = 3

Practice 1

Find each quotient.

6,300 ÷ 700 = _____

6,300 ÷ 70 = _____

6,300 ÷ 7 = _____

Basic fact:
63 ÷ 7 = _____

Example 2

Find the quotients.

3,000 ÷ 6 = 500

3,000 ÷ 60 = 50

3,000 ÷ 600 = 5

Practice 2

Find the quotients.

4,000 ÷ 500 = _____

4,000 ÷ 50 = _____

4,000 ÷ 5 = _____

On Your Own

Find each quotient. Give the basic fact for each one.

a. 3,600 ÷ 4 = _____

   basic fact: _____

b. 5,600 ÷ 70 = _____

   basic fact: _____

c. 2,000 ÷ 400 = _____

   basic fact: _____

Write About It

What is wrong with this solution?

2,000 ÷ 50 = ?

The basic fact is 20 ÷ 5 = 4.

    3 zeros in the dividend
  − 1 zero in the divisor
  ─────────────────────
    2 zeros in the quotient

2,000 ÷ 50 = 400

Objective 8.1: Solve division problems with multiples of 10.

Volume 3 — 42 — Level F

Lesson 8-1 — **Divide Multiples of 10** — Try It

1. Find the quotients.

   6 ÷ 2 = _____

   60 ÷ 2 = _____

   600 ÷ 2 = _____

   6,000 ÷ 2 = _____

2. Find each quotient.

   Basic fact: 72 ÷ 9 = _____

   7,200 ÷ 9 = _____

   7,200 ÷ 90 = _____

   7,200 ÷ 900 = _____

3. What is 4,000 ÷ 50? Circle the letter of the correct answer.

   A 8   **B** 80

   C 800   **D** 8,000

4. A roll of ribbon is 300 centimeters long. Katie cuts the ribbon into 10 equal pieces. How long is each piece?

   300 ÷ 10 = _____

   Each piece is _____ cm long.

5. Solve. Write the basic fact for each one.

   a. 270 ÷ 3 = _____

   basic fact: _____

   b. 6,400 ÷ 800 = _____

   basic fact: _____

   c. 35,000 ÷ 500 = _____

   basic fact: _____

6. Which is **not** equal to 8? Circle the letter of the correct answer.

   A 4,000 ÷ 50   **B** 4,000 ÷ 500

   C 40 ÷ 5   **D** 400 ÷ 50

7. Match each problem to its quotient.

   1,600 ÷ 400   400

   800 ÷ 20   40

   2,400 ÷ 6   4

8. Explain how to divide 3,000 ÷ 5.

Objective 8.1: Solve division problems with multiples of 10.

Volume 3 — 43 — Level F

Lesson 8-2: Estimate Quotients

Model It — A

Words to Know An **estimate** is a number close to the exact answer.

Activity 1

Estimate 8,753 ÷ 3.

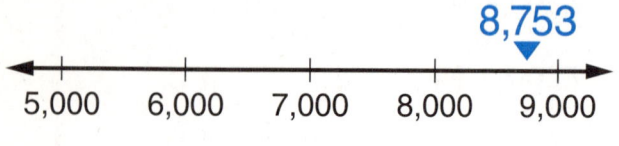

```
          8,753
            ▼
  |----+----+----+----+----|
5,000 6,000 7,000 8,000 9,000
```

8,753 rounds to 9,000.

Think: 9,000 ÷ 3 = 3,000.

8,753 ÷ 3 is about 3,000.

Practice 1

Estimate 367 ÷ 4.

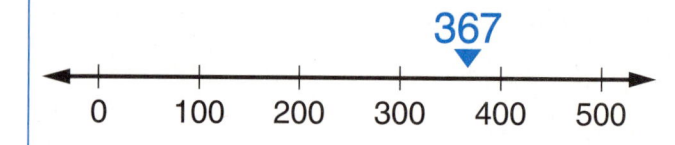

```
            367
             ▼
  |----+----+----+----+----|
  0   100  200  300  400  500
```

367 rounds to _____.

Think: _____ ÷ 4 = _____.

367 ÷ 4 is about _____.

Activity 2

Estimate 3,629 ÷ 18.

3,629 rounds to 4,000.
18 rounds to 20.

3,629 ÷ 18 → 4,000 ÷ 20 = 200

3,629 ÷ 18 ≈ 200

Practice 2

Estimate 8,754 ÷ 29.

8,754 rounds to _____.

29 rounds to _____.

8,754 ÷ 29 → _____ ÷ _____ = _____

8,754 ÷ 29 ≈ _____

On Your Own

Estimate each quotient.

a. 816   ÷   234
    ↓   ↓

    _____ ÷ _____ = _____

b. 5,814 ÷   322
    ↓   ↓

    _____ ÷ _____ = _____

Write About It

Write out the steps that you would use when estimating a quotient.

Objective 8.2: Estimate quotients by rounding numbers and using mental math techniques.

Lesson 8-2: Estimate Quotients

B Understand It

Words to Know Compatible numbers are close to the actual numbers but easier to work with mentally.

Example 1

Use compatible numbers to estimate.

5,637 ÷ 9

56 ÷ 9 is **not** a basic fact.
54 is near 56 and **is** a multiple of 9.

5,637 ÷ 9
↓ ↓
5,400 ÷ 9 = 600

5,637 ÷ 9 = about 600

Practice 1

Use compatible numbers to estimate.

6,425 ÷ 7

Is 64 ÷ 7 a basic fact? _____
What number is near 64 **and** is a multiple of 7? _____

6,425 ÷ 7
↓ ↓
_____ ÷ 7 = _____

6,425 ÷ 7 = about _____

Example 2

Estimate. Round the divisor first, and then choose a compatible dividend.

2,484 ÷ 745
↓
2,484 ÷ 700
↓
2,100 ÷ 700 = 3

Think: 21 ÷ 7 is a basic fact.
2,484 is close to 2,100.
Use 2,100.

So 2,484 ÷ 745 ≈ 3.

Practice 2

Estimate 4,524 ÷ 75.

75 rounds to _____.

What basic fact can you use? _____

4,524 ÷ 75
↓ ↓
_____ ÷ _____ = _____

So 4,524 ÷ 75 ≈ _____.

On Your Own

Use compatible numbers to estimate.

a. 4,124 ÷ 35
   ↓   ↓
   _____ ÷ _____ = _____

b. 524 ÷ 65
   ↓   ↓
   _____ ÷ _____ = _____

Write About It

When is it better to use compatible numbers? When is it better to round?

Objective 8.2: Estimate quotients by rounding numbers and using mental math techniques.

Volume 3 — Level F

Lesson 8-2: Estimate Quotients

Try It — C

1. Round to estimate 284 ÷ 5.

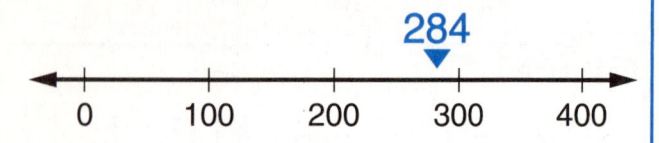

[number line from 0 to 400 marked at 100, 200, 300, 400 with 284 indicated]

284 rounds to _____.

Think: _____ ÷ 5 = _____

284 ÷ 5 = about _____

2. Use compatible numbers to estimate.

5,513 ÷ 6

What basic fact can you use to round the dividend? _____

5,513 ÷ 6
  ↓   ↓
_____ ÷ 6 = _____

5,513 ÷ 6 ≈ _____

3. For the **best** estimate of 3,319 ÷ 80, which number would you choose as a compatible number for the dividend? Circle the letter of the correct answer.

A 3,000   B 3,200

C 3,300   D 4,000

4. Round to estimate 6,224 ÷ 317.

6,224 rounds to _____.

317 rounds to _____.

So, 6,224 ÷ 317 ≈ _____

5. Use compatible numbers to estimate.

3,020 ÷ 795
  ↓   ↓
_____ ÷ _____ = _____

6. There are 238 people in line for a roller coaster. If each car of the roller coaster can hold 4 people, about how many cars will be filled by the people in line?

_____ ÷ _____ ≈ _____

7. Round to estimate 3,247 ÷ 644. Circle the letter of the correct answer.

A 4   B 5

C 6   D 50

8. Alexa and Robin both use compatible numbers to estimate the quotient of 4,168 ÷ 519. Alexa rounds the problem to 4,000 ÷ 500. Robin rounds the problem to 4,500 ÷ 500. Whose estimate will be closest to the exact answer? How do you know?

Objective 8.2: Estimate quotients by rounding numbers and using mental math techniques.

Lesson 8-3 Divide by 1-Digit Numbers

A Model It

Words to Know A **remainder** is an amount left over when one number is divided by another.

Activity 1

Use base ten blocks to divide: 963 ÷ 3.

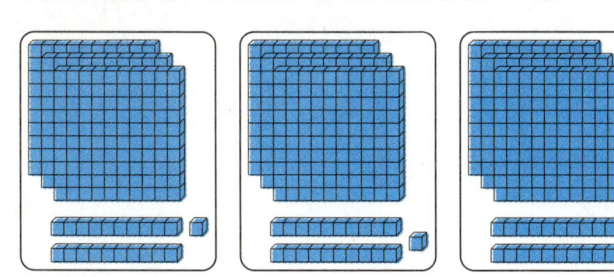

Start with 963. Separate the blocks equally among 3 groups.

963 ÷ 3 = 321

Practice 1

Use base ten blocks to divide: 246 ÷ 2.

Start with 246. Separate the blocks equally between 2 groups.

246 ÷ 2 = _____

Activity 2

Use base ten blocks to divide 98 by 3.

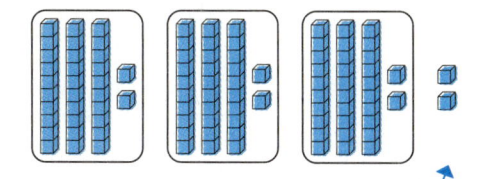

98 ÷ 3 = 32 R2

These can't be evenly divided. This is the remainder.

Practice 2

Use base ten blocks to divide 50 by 4.

50 ÷ 4 = _____ R _____

On Your Own

Use base ten blocks to divide: 863 ÷ 2.

863 ÷ 2 = _____

Write About It

When will a division problem have a remainder?

Objective 8.3: Divide multidigit numbers by 1-digit numbers with and without remainders.

Volume 3 — 47 — Level F

Lesson 8-3 **Divide by 1-Digit Numbers**

B Understand It

Example 1

Use paper and pencil to find 93 ÷ 4.

```
   23 R1
4)93
  -8
   13
  -12
    1
```

Start with 9 tens ÷ 4. There will be 2 tens in each group, so write 2 in the quotient. Multiply, subtract, bring down. Next find 13 ones ÷ 4. When there is nothing more to bring down, write the remainder.

93 ÷ 4 = 23 R1

Practice 1

Use paper and pencil to find 329 ÷ 5.

329 ÷ 5 = _____

Example 2

Use paper and pencil to find 326 ÷ 5.

```
   25 R1
5)326
  -30
   26
  -25
    1
```

326 ÷ 5 = 65 R1

Check: Multiply the quotient by the divisor.
65 × 5 = 325

Add the remainder.
325 + 1 = 326

Practice 2

Use paper and pencil to find 347 ÷ 8.

Check: Multiply and add.

_____ × 8 = _____

_____ + _____ = _____

347 ÷ 8 = _____

On Your Own

3,548 ÷ 4 = _____

Write About It

How would you check your answer to On Your Own?

Objective 8.3: Divide multidigit numbers by 1-digit numbers with and without remainders.

Lesson 8-3: Divide by 1-Digit Numbers

Try It C

1. Use paper and pencil to find 85 ÷ 4.

2. Use paper and pencil to find 5,244 ÷ 6.

3. Which problem does the model show?

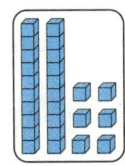

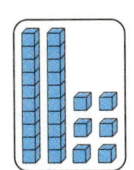

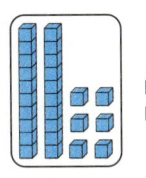

   A 80 ÷ 2   **B** 80 ÷ 3

   C 80 ÷ 4   **D** 26 ÷ 3

4. Which problem has a remainder?

   A 525 ÷ 5   **B** 612 ÷ 3

   C 452 ÷ 6   **D** 432 ÷ 8

5. Steven baked 50 energy bars. He wants to put them in bags of 6 bars each and freeze them. How many bags can he make? How many bars, if any, will be left over?

   _____

   _____

6. Which division is correct for 585 ÷ 3?

   ```
         195          160         168 R1       193 R1
   A 3)585      B 3)585      C 3)585      D 3)585
       -3           -3           -3           -3
       28           18           20           28
      -27          -18          -18          -27
       15            0           25           10
      -15                       -24           -9
        0                         1            1
   ```

7. Use base ten blocks to divide 139 ÷ 4.

   139 ÷ 4 = _____

8. Erin wants to divide 83 photos evenly on the 11 pages of her scrapbook. How many photos can she fit on each page? How many photos will be left over?

   _____

   _____

Objective 8.3: Divide multidigit numbers by 1-digit numbers with and without remainders.

Volume 3 — Level F

Lesson 8-4: Choose a Method

Model It — A

Activity 1

Estimate. 624 ÷ 4 = _____

1. Round the dividend. 624 → 600
2. Look for a basic fact in 600 ÷ 4. Neither 6 ÷ 4 nor 60 ÷ 4 is a basic fact.
3. 60 ÷ 4 divides evenly. How many fours are in 60? There are 15 fours in 60, so 60 ÷ 4 = 15 is a fact.
4. Divide. 600 ÷ 4 = 150
   624 ÷ 4 ≈ 150

Practice 1

Estimate. 4,942 ÷ 2 = _____

1. Round the dividend. 4,942 → _____
2. Remove some zeros to look for a simpler division fact you can solve.
3. What is the simpler division fact? _____
4. Divide. _____
   4,942 ÷ 2 ≈ _____

Activity 2

3,178 games will be stored on 39 racks. **About** how many games should go on each rack?

1. Round both numbers. 3,178 → 3,000
   39 → 40
2. Look for a basic fact in 3,000 ÷ 40. Neither 3 ÷ 4 nor 30 ÷ 4 will divide evenly.
3. Look for a number close to 3,178 that works with 40. Choosing 3,200 ÷ 40, we see the basic fact 32 ÷ 4 = 8.
4. Divide. 3,200 ÷ 40 = 80

About 80 games should go on each rack.

Practice 2

Thuy's Diner served 4,024 meals in 63 days. **Approximately** how many meals per day did the diner serve?

1. Round both numbers. 4,024 → _____
   63 → _____
2. If you find a basic fact, divide. If not, find a compatible dividend. If you use _____, you can see the basic fact _____.
3. Divide. _____
   4,024 ÷ 63 ≈ _____

The diner served about _____ meals per day.

On Your Own

Rafael needs a rough idea of how many meeting rooms he will need. He thinks that 4,600 people will attend the conference, and knows he can fit 370 people in each room. Solve.

_____

Write About It

Here are two estimates of 43,500 ÷ 9. Which is closer to the exact answer? Why?

45,000 ÷ 9 = 5,000	36,000 ÷ 9 = 4,000

Objective 8.4: Solve division problems by choosing between mental math, pencil and paper, or estimation.

Lesson 8-4 Choose a Method

B Understand It

Example 1

Betty has to put 207 balloons in 9 equal bunches. How many balloons should be in each bunch?
You need the exact answer.
There are no basic facts in 207 ÷ 9.
Use paper and pencil.

```
      32
   9)207
    −18
     27
    −27
      0
```

207 ÷ 9 = 23 balloons per bunch

Practice 1

A pet store owner has to divide a shipment of 544 goldfish evenly among 8 tanks. How many goldfish will go in each tank?

Do you need the exact answer? _____

Do you see a basic fact? _____
Use paper and pencil.

544 ÷ 8 = _____ goldfish per tank

Example 2

Lauren typed 630 words in 9 minutes. What was her speed in words per minute?

  typing speed = total words ÷ time

You need the exact answer.
There is a basic fact in 630 ÷ 9.
Use 63 ÷ 9 = 7 to divide mentally.
If 63 ÷ 9 = 7, then 630 ÷ 9 = 70.
Her speed was 70 words per minute.

Practice 2

A conservation team planted 840 trees in 12 hours. What was their speed in trees planted per hour?

Do you need the exact answer? _____

Do you see a basic fact? _____

Use the basic fact _____ ÷ _____ = _____
to divide mentally.

840 ÷ 12 = _____

The team planted _____ trees per hour.

On Your Own

Circle the problem you can use mental math to solve. Then solve both.

2,800 ÷ 70 = _____   2,730 ÷ 70 = _____

Write About It

What two questions should you ask yourself before choosing a method for solving a division word problem?

Objective 8.4: Solve division problems by choosing between mental math, pencil and paper, or estimation.

Lesson 8-4 **Choose a Method** **Try It**

1. Estimate 6,218 ÷ 68.

   6,218 ÷ 68 ≈ _____

2. Which calculation will provide an estimate that is closest to the exact answer of 4,187 ÷ 86? Circle the letter of the correct answer.

   A 4,000 ÷ 100 = 40   **B** 4,500 ÷ 90 = 50

   C 4,800 ÷ 80 = 60   **D** 4,200 ÷ 60 = 70

3. Julie earned $320 for working 40 hours. How much did Julie earn per hour?

   _____

4. A school rented 12 buses to take 238 students on a field trip. About how many students were on each bus?

   About _____ students

5. Which is greater: 158 ÷ 6, or 248 ÷ 18?

   _____

   How did you decide?

6. A box of beads contains 2,382 beads. If Sarah wants to make 78 bracelets using the beads, about how many beads will be on each bracelet?

   About _____ beads

7. Find an exact answer for 1,208 ÷ 8.

   1,208 ÷ 8 = _____

   Use estimation to show that your answer is reasonable.

8. A florist has 540 roses. He plans to make bouquets using 6 roses in each bouquet. How many bouquets can he make?

   a. Do you need an exact answer or an estimate? _____

   b. What is the basic fact you can use?
   _____

   c. How many bouquets can he make?
   _____ bouquets

Objective 8.4: Solve division problems by choosing between mental math, pencil and paper, or estimation.

Lesson 8-5: Multiply and Divide

Model It

Activity 1

A box holds 6 muffins. If the bakery sold 32 boxes of muffins, about how many muffins were sold?
Here are two of the boxes:

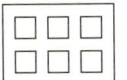

The picture shows combining, so multiply.
About means to find an estimate.
32 rounds to 30, and 30 × 6 = 180.

The bakery sold about 180 muffins.

Practice 1

Tyson feeds his dogs 5 cups of kibble each day. About how much kibble does he need to bring with him when he takes the dogs on an 18-day trip?
Draw to show the kibble for 2 days.

Separate or combine? _____

Estimate or exact answer? _____

He needs about _____ cups of kibble.

Activity 2

164 pastels must be put back in 4 boxes. How many pastels belong in each box?
Start with 16 pastels and 4 boxes:

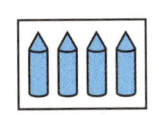

The picture shows separating, so divide.
Find the exact answer. 164 ÷ 4 = 41
41 pastels belong in each box.

Practice 2

You want to tie equal bunches of balloons on 18 carnival booths. If you have 216 balloons, how many can be in each bunch?
Start with 3 booths and 9 balloons. Draw.

Separate or combine? _____

Estimate or exact answer? _____

You can tie _____ balloons in each bunch.

On Your Own

762 bricks were used to build a patio. They were arranged in 19 rows. About how many bricks are in each row?

_____

Write About It

Write a word problem for 35 × 7.

Objective 8.5: Solve word problems using either multiplication or division.

Lesson 8-5: Multiply and Divide

B — Understand It

Example 1

The principal assigned 23 students to each class. If there are 6 classes, how many students are there in all?

Solve a simpler problem. If there are 10 students in each class and only 2 classes, there would be 20 students. Multiply.

Find the exact answer.

```
  1
  23
× 6
 138
```

There are 138 students in all.

Practice 1

Tina is reading a 231-page book. She reads the same number of pages each day. How many pages does she need to read each day to finish the book in 7 days?

Solve a simpler problem. If there are 21 pages to finish in 7 days, how many pages must she read per day? _____ Multiply or divide?

_____.

Tina needs to read

_____ pages each day

Example 2

The coach purchased 30 baseball caps. The total cost was $238. What was the approximate cost of each baseball cap?

Solve a simpler problem. If there are 3 caps and the total cost is $30, each cap cost $10. Divide.
Approximate tells us to estimate.
$238 \div 30 \rightarrow 240 \div 30 = 8$

Each baseball cap cost about $8.

Practice 2

The fundraisers asked each donor to give $17. If 42 donors gave that amount, about how much money was raised?

Solve a simpler problem. If 4 donors each gave $1, how much was raised? _____
Divide or multiply? _____.
Circle the word that tells you to estimate.
About _____ was raised.

On Your Own

Brad is grilling 52 veggie skewers for the picnic. If there are 12 vegetable pieces on each skewer, how many vegetable pieces should his helper provide?

_____

Write About It

Write a word problem that $144 \div 24$ can be used to solve.

Objective 8.5: Solve word problems by using either multiplication or division.

Lesson 8-5 — **Multiply and Divide** — Try It

1. A group of 7 people went to the science museum. The total cost for the tickets was $112. What was the cost per ticket?

   The cost was $ _____ per ticket.

2. There are 52 weeks in a year. How many weeks are there in 5 years?

   There are _____ weeks in 5 years.

3. Which fact could be used to solve the following problem? Circle the letter of the correct answer.

   There are 36 plastic cups in a bag. How many cups are in 9 bags?

   A 36 + 9   **B** 36 − 9

   C 36 × 9   **D** 36 ÷ 9

4. You are given the total number of seats in an auditorium and the number of rows. What operation is used to find the number of seats in each row? Circle the letter of the correct answer.

   A multiplication   **B** division

   C addition   **D** subtraction

5. Which of the following problems could use 15 × 5 to solve it? Circle the letter of the correct answer.

   A A DVD costs $15. How much will 5 DVDs cost?

   B $15 is spent on 5 DVDs. What is the cost of each DVD?

   C A DVD costs $15 and is on sale for $5 off. What is the sale price of the DVD?

   D Several DVDs totaling $15 were purchased online. The shipping cost is $5. What is the total cost of the DVDs?

6. The school cafeteria receives 12 boxes of glasses. Each box has 48 glasses. How many glasses are there? Write a fact that could be solved to answer the question. Explain the chosen operation.

7. Estimate the solution to Exercise 6.

   There are approximately _____ glasses.

8. A girl has 76 photos to put into an album. She can put 4 photos on a page. How many pages will she need?

   She will need _____ pages.

Objective 8.5: Solve word problems by using either multiplication or division.

Volume 3   Level F

Topic 8: Use Division to Compute

Topic Summary

Choose the letter of the correct answer. Explain how you decided.

1. What is 600,000 ÷ 200?

   A 30

   B 300

   C 3,000

   D 30,000

   _____

   _____

   _____

2. Arianna's photo album holds 288 photographs. She can fit 4 photographs on each page. How many pages are in the photo album?

   A 72

   B 284

   C 292

   D 1,152

   _____

   _____

   _____

Objective: Review skills related to division.

Topic 8

Use Division to Compute

Mixed Review

1. A clothing store sold 137 pairs of jeans in a month. If they sell the same amount each month, how many pairs will they sell in 6 months?

A 143

B 682

C 792

D 82

Volume 3, Lesson 7-3

2. At a dance recital, there are 28 dancers on the stage and 25 more in the dressing rooms. How many dancers are at the recital? Circle the letter of the correct answer.

A 43

B 53

C 55

D 60

Volume 3, Lesson 6-4

3. Order 59, 915, 519, and 95 from greatest to least.

_____

Volume 1, Lesson 1-4

4. Collin estimated 8,426 ÷ 9 as 8,000 ÷ 9. What would be a better estimate?

Volume 3, Lesson 8-2

5. Write the place of each underlined digit.

a. 2,3<u>1</u>2 _____

b. 4,<u>1</u>31 _____

c. 3<u>0</u>,000 _____

d. 99<u>9</u> _____

Volume 1, Lesson 2-2

6. Find each quotient.

a. 56 ÷ 8 = _____

b. 36 ÷ 6 = _____

c. 27 ÷ 3 = _____

d. 32 ÷ 4 = _____

Volume 1, Lesson 5-2

7. Use the make-ten strategy to find the sum for 8 + 6.

Volume 2, Lesson 3-2

8. Subtract 100 − 61.

_____

Volume 2, Lesson 6-6

Objective: Maintain concepts and skills.

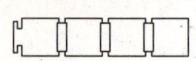

Topic 9

Basic Properties of Algebra

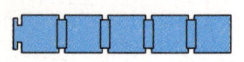

Topic Introduction

Complete with teacher help if needed.

1. Write two addition sentences about the picture.

   ____ + ____ = ____

   ____ + ____ = ____

   Objective 9.1: Use the commutative and associative properties of addition.

2. Simplify.

   a. $3(2 + 4) =$ _____

   b. $3(2 + 5) =$ _____

   c. $5(6 - 3) =$ _____

   Objective 9.3: Know and use the distributive property.

3. Use the commutative property to write the expression another way.

   $3 \times 4 = 12$

   ____ × ____ = ____

   Objective 9.2: Recognize and use the commutative and associative properties of multiplication.

4. Find $5 \times 8 \times 2$.

   a. One way:

   $(8 \times 5) \times 2 =$ _____ $\times 2$

   $=$ _____

   b. Another way:

   $8 \times ($ _____ $\times$ _____ $) = 8 \times$ _____

   $=$ _____

   c. This shows the _____ property.

   Objective 9.2: Recognize and use the commutative and associative properties of multiplication.

Volume 3 — 58 — Level F

Lesson 9-1: Properties of Addition

Model It

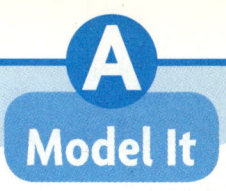

Words to Know
Commutative property of addition: You can add two numbers in any order and the sum will be the same. For example, 6 + 5 = 11 and 5 + 6 = 11.
Associative property of addition: You can group numbers in any way and the sum will still be the same. For example, (2 + 8) + 7 = 17 and 2 + (8 + 7) = 17.

Activity 1

Use MathFlaps to show 7 + 5 = 5 + 7.

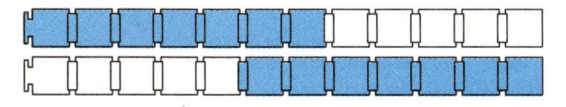

7 + 5 = 12   5 + 7 = 12

There are 12 MathFlaps in each model. The sum is the same when the numbers are added in a different order.

Practice 1

Use MathFlaps to show 2 + 4 = 4 + 2.

____ + ____ = ____   ____ + ____ = ____

There are ____ MathFlaps in each model. The sum is the same.

Activity 2

Use MathFlaps to show
(2 + 3) + 4 = 2 + (3 + 4).

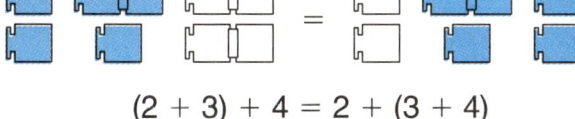

(2 + 3) + 4 = 2 + (3 + 4)
    5   + 4 = 2 +   7
         9   =   9

The sum was the same when the numbers were grouped differently to add.

Practice 2

Use MathFlaps to show
(2 + 5) + 1 = 2 + (5 + 1).

(____ + ____) + ____ = ____ + (____ + ____)

____ + 1 = 2 + ____

____ = ____

On Your Own

Write the missing number in each equation.

(40 + 8) + 2 = 40 + (____ + 2)

15 + ____ = 16 + 15

Write About It

Use mental math and the associative property to solve 78 + 6 + 4. Explain your work.

Objective 9.1: Use the commutative and associative properties of addition.

Lesson 9-1: Properties of Addition

Example 1

The MathFlaps show 9 + 8 = 17.

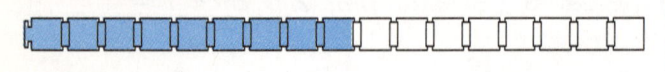

9 + 8 = 17

They also show 8 + 9 = 17.

8 + 9 = 17

Both models show 17 MathFlaps, so, 9 + 8 = 8 + 9.

Practice 1

Write the number sentence shown.

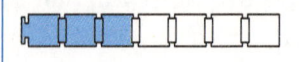

____ + ____ = ____

Then write another sentence using the same MathFlaps.

____ + ____ = ____

Both models show ____ MathFlaps,

so, 3 + ____ = ____ + ____.

Example 2

The properties of addition can be used to add mentally.

16 + 27 + 4
= 27 + 16 + 4   commutative property
= 27 + 20   associative property
= 47

Practice 2

Use the properties of addition to add.

15 + 48 + 5

= 15 + ___ + 48   _____ property

= ____ + 48   _____ property

= ____

On Your Own

Use the properties to add. Show your work.

12 + 9 + 8 + 11

Write About It

Show an example of the commutative property and an example of the associative property. Label each example. Use parentheses in one of the examples, and explain why the parentheses are helpful.

Objective 9.1: Use the commutative and associative properties of addition.

Lesson 9-1 **Properties of Addition** Try It

1. Shade the MathFlaps in two ways to show that 2 + 6 = 6 + 2. Write the fact for each picture.

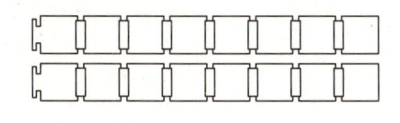

   _____ + _____ = _____

   _____ + _____ = _____

2. Write the missing number.

   a. 8 + 1 = _____ + 8

   b. _____ + (6 + 10) = (3 + 6) + 10

   c. _____ + 19 = 19 + 57

   d. (10 + _____) + 3 = 10 + (25 + 3)

3. Identify the property used. Write C for commutative, A for associative, or B for both.

   a. 1 + 4 = 4 + 1 _____

   b. (4 + 5) + 9 = (5 + 4) + 9 _____

   c. (2 + 8) + 1 = 2 + (8 + 1) _____

   d. 7 + 5 + 3 = (7 + 3) + 5 _____

4. Which sentence shows the associative property? Circle the letter of the correct answer.

   A 11 + 39 = 39 + 11

   B (28 + 96) = (96 + 28)

   C 50 + (30 + 6) = 50 + (6 + 30)

   D (9 + 81) + 1 = 9 + (81 + 1)

5. Use the properties to add. Show your work.
   2 + 4 + 8 + 6

6. Explain how you can use the associative and commutative properties to find the sum.
   0.2 + 0.4 + 0.8 + 0.6

7. Add from left to right. Then use the properties to check your work.
   46 + 92 + 25 + 8 + 4

8. Explain why thinking of 204 as 200 + 4 would make it easier to use the properties to add 146 + 204.

Objective 9.1: Use the commutative and associative properties of addition.

Volume 3 61 Level F

Lesson 9-2: Properties of Multiplication

Model It — A

Words to Know

Commutative property of multiplication: You can multiply two numbers in any order and the product will be the same. For example, $5 \times 6 = 30$ and $6 \times 5 = 30$.

Associative property of multiplication: You can group numbers in any way and the product will still be the same. For example, $(2 \times 7) \times 8 = 112$ and $2 \times (7 \times 8) = 112$.

Activity 1

Draw a diagram to show $5 \times 2 = 2 \times 5$.

♥♥
♥♥
♥♥   ♥♥♥♥♥
♥♥   ♥♥♥♥♥
♥♥

$5 \times 2 = 10$ $2 \times 5 = 10$
So, $5 \times 2 = 2 \times 5$.

Practice 1

Draw a picture to show that $3 \times 4 = 4 \times 3$.

So, $3 \times 4 = $ _____ $\times$ _____ .

Activity 2

Use the commutative and associative properties to find $2 \times 17 \times 5$.

$2 \times 17 \times 5$

$= 2 \times 5 \times 17$ commutative property

$= (2 \times 5) \times 17$ associative property

$= 10 \times 17 = 170$

Practice 2

Use the commutative and associative properties to find $5 \times 11 \times 2$.

$5 \times 11 \times 2$

$= 5 \times $ _____ $\times 11$ _____

$= ($ _____ $\times$ _____ $) \times 11$ _____

$= $ _____ $\times 11 = $ _____

On Your Own

Write the missing number in each equation.

$8 \times 9 = 9 \times $ _____

$(6 \times 8) \times 7 = $ _____ $\times (8 \times 7)$

Write About It

Ariel said that $3 \times 5 = 5 + 3$ is an example of the commutative property. Do you agree? Explain.

Objective 9.2: Recognize and use the commutative and associative properties of multiplication.

Lesson 9-2: Properties of Multiplication

Understand It — B

Example 1

If $3 \times 21 \times 8 = 504$, then what is the value of $3 \times 8 \times 21$?

Use the commutative property to change the order of the factors.
$3 \times 8 \times 21 = 3 \times 21 \times 8$

Since the two expressions are equal, their products would also be equal.
$3 \times 8 \times 21 = 504$

Practice 1

If $8 \times 12 \times 2 = 192$, then what is the value of $12 \times 8 \times 2$?

Use the _____ property.

$12 \times 8 \times 2 =$ _____ $\times$ _____ $\times 2$

Since the two expressions are equal, their products would also be equal.

$12 \times 8 \times 2 =$ _____

Example 2

Use the properties of multiplication to simplify multiplication.

$$5 \times 8 \times 2 \times 3$$

$= 5 \times 2 \times 8 \times 3$ commutative property

$= (5 \times 2) \times (8 \times 3)$ associative property

$= 10 \times 24$ Simplify.

$= 240$

Practice 2

Use the properties of multiplication to simplify multiplication.

$$4 \times 3 \times 5 \times 7$$

$= 4 \times$ _____ $\times 3 \times 7$ _____ property

$= ($ _____ $) \times (3 \times 7)$ _____ property

$=$ _____ $\times$ _____

$=$ _____

On Your Own

Use the properties to multiply.

$6 \times 3 \times 5 \times 5$

$9 \times 4 \times 1 \times 2$

Write About It

Explain how you know that 10×162 is equal to $2 \times 162 \times 5$.

Objective 9.2: Recognize and use the commutative and associative properties of multiplication.

Volume 3   Level F

Lesson 9-2: Properties of Multiplication

Try It — C

1. Draw a picture to show that 4 × 9 = 9 × 4.

2. Write each missing number.

   a. 2 × (_____ × 3) = (2 × 6) × 3

   b. 7 × 13 = _____ × 7

   c. 9 × (3 × 7) = (_____ × 3) × 7

   d. 19 × _____ = 32 × 19

   e. 20 × (_____ × 4) = 20 × (4 × 6)

3. Identify the property used. Write C for commutative, A for associative, or B for both.

   a. 13 × (18 × 53) = (13 × 18) × 53 _____

   b. (12 × 8) × 15 = 8 × (12 × 15) _____

   c. 19 × (2 × 8) = (19 × 2) × 8 _____

   d. 28 × 11 = 11 × 28 _____

4. Which sentence shows the commutative property of multiplication? Circle the letter of the correct answer.

   A 12 + 6 = 6 + 12

   B 9 × (5 × 8) = (9 × 5) × 8

   C 16 × (12 × 11) = 16 × (11 × 12)

   D (16 × 5) × 3 = 16 × (5 × 3)

5. Give an example of an everyday action that is commutative.

6. Multiply from left to right. Then use the properties to check your work.

   14 × 4 × 2 × 25 =

7. If 6 × 23 × 8 = 1,104, how can you find the product of 23 × 8 × 6 without multiplying?

8. Use the commutative and associative properties to simplify the product of the numbers 1 through 5.

Objective 9.2: Recognize and use the commutative and associative properties of multiplication.

Lesson 9-3: The Distributive Property

Model It — A

Words to Know: The **distributive property** means multiplying a sum (or difference) by a number gives the same result as multiplying each number in the sum (or difference) by the number and adding (or subtracting) the products. For example, $8 \times (2 + 7) = (8 \times 2) + (8 \times 7)$.

Activity 1

Use the distributive property to multiply $4(6 + 5)$.

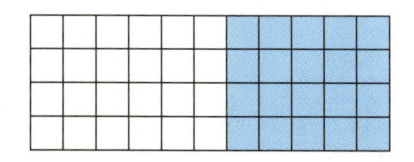

$(4 \times 6) + (4 \times 5)$

$24 + 20 = 44$

$4 \times 11 = (4 \times 6) + (4 \times 5) = 24 + 20 = 44$

Practice 1

Use the distributive property to multiply $5(10 + 3)$

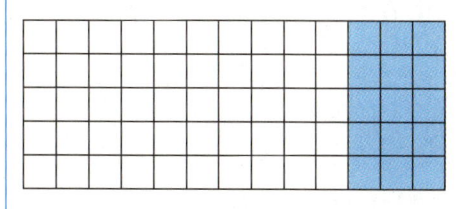

$5 \times (10 + 3)$

$= (5 \times \underline{}) + (5 \times \underline{})$

$= \underline{} + \underline{} = \underline{}$

Activity 2

Use the distributive property to find $3(2 + 6)$.

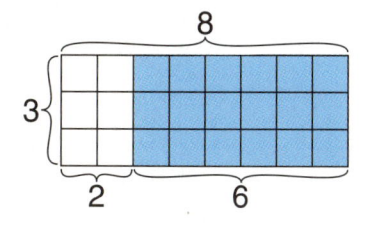

$(3 \times 2) + (3 \times 6)$

$6 + 18 = 24$

Check: $3(2 + 6) = 3 \times 8 = 24$

Practice 2

Draw and shade a rectangle to show how to use the distributive property to find $4(3 + 9)$.

$4(3 + 9) = (4 \times \underline{}) + (4 \times \underline{})$

$= \underline{} + \underline{} = \underline{}$

Check: $4(3 + 9) = 4 \times 12 = \underline{}$

On Your Own

Use a rectangle to illustrate using the distributive property to solve $5(12 + 4)$.

Write About It

Explain how to use the distributive property to solve $8(7 + 3)$.

Objective 9.3: Know and use the distributive property.

Lesson 9-3: The Distributive Property

B Understand It

Example

Since Jamal didn't know the product of 7×15, he wrote the greater number, 15, as a **sum**:

$7 \times 15 = 7 \times (9 + 6)$.

Then he used the distributive property:

$(7 \times 9) + (7 \times 6) = 63 + 42 = 105$, so $7 \times 15 = 105$.

Then he checked by writing 15 as a **difference**:

$7 \times 15 = 7 \times (20 - 5)$.

Then he used the distributive property:

$(7 \times 20) - (7 \times 5) = 140 - 35 = 105$, so $7 \times 15 = 105$.

Practice

Show 8×19 as $8 \times$ (a sum equal to 19). Solve.

$8 \times 19 = 8 \times ($ _____ $)$

$ = (8 \times $ _____ $) + (8 \times $ _____ $)$

$ = $ _____

Show 8×19 as $8 \times$ (a difference equal to 19). Solve.

$8 \times 19 = 8 \times ($ _____ $)$

$ = (8 \times $ _____ $) - (8 \times $ _____ $)$

$ = $ _____

On Your Own

Choose a problem that will help you solve 3×14. Show how to use the distributive property to solve.

$3 \times (10 \times 4)$ $\quad$ $3 \times (10 + 4)$

$3 \times (15 - 1)$ $\quad$ $3 \times (10 - 4)$

Write About It

Explain how to use the distributive property to find 6×23 mentally.

Objective 9.3: Know and use the distributive property.

Lesson 9-3 — **The Distributive Property** — **Try It**

1. Use the distributive property to find 7(3 + 8). Shade the picture to help.

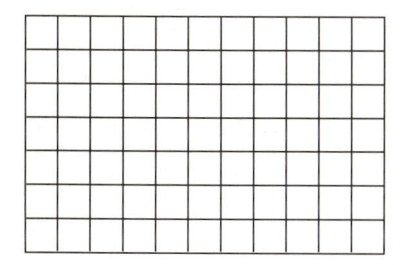

   _____

2. Use the distributive property to rewrite each expression. Then find each product.

   a. 9(2 + 6) =

   _____

   b. 2(5 + 4) =

   _____

   c. 3(12 − 4) =

   _____

3. Ryan said that 31 × 45 = 1,305. Do you agree? Use the distributive property to explain.

4. Which expression is equivalent to 99 × 52? Circle the letter of the correct answer.

   A (100 × 52) − (1 × 52)

   B (100 × 52) + (1 × 52)

   C (99 × 50) − (9 × 2)

   D (99 × 50) + (9 × 2)

5. How can you use the distributive property to find 101 × 15?

6. How can you use the distributive property to find 18 × 22?

Objective 9.3: Know and use the distributive property.

Topic 9: Basic Properties of Algebra

Topic Summary

Choose the correct answer. Explain how you decided.

1. What is the sum of 16 + 21 + 4 + 9? Use the associative and commutative properties.

    A 40

    B 49

    C 50

    D 60

    _____
    _____
    _____

2. Simplify 6(10 − 8). Use the properties to help.

    A 52

    B 12

    C 60

    D 48

    _____
    _____
    _____

Objective: Review the associative, commutative, and distributive properties.

Topic 9

Basic Properties of Algebra

Mixed Review

1. Which number is the missing number in the equation 8 + ☐ = 17. Circle the letter of the correct answer.

   A 6   B 7

   C 8   D 9

   Volume 2, Lesson 3-2

2. Multiply.

   a. 19 × 4 = _____

   b. 46 × 7 = _____

   c. 63 × 2 = _____

   d. 108 × 9 = _____

   Volume 3, Lesson 7-3

3. Find the missing number.

   a. _____ + 30 = 30 + 29

   b. (4 + 1) + 7 = 4 + (1 + _____)

   c. 9 + (_____ + 5) = (9 + 3) + 5

   d. 3 + 4 + 6 = (3 + _____) + 4

   Volume 3, Lesson 9-1

4. Write the number in standard form.

   a. 6 hundreds, 2 tens _____

   b. 4 thousands, 5 ones _____

   c. 5 hundreds, 3 tens, 9 ones _____

   d. 2 thousands, 1 hundred _____

   Volume 1, Lesson 2-1

5. Which number is divisible by 6?

   A 317

   B 518

   C 702

   D 1,245

   Volume 2, Lesson 5-3

6. Write the number shown by the model.

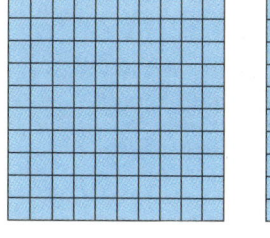

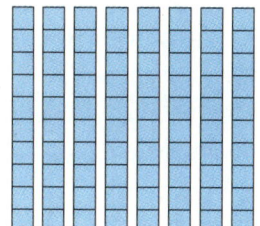

   _____

   Volume 1, Lesson 1-2

7. Manny had 37 trading cards. He traded 18 cards with his friends. How many cards does he have left?

   _____

   Volume 3, Lesson 6-5

8. Order 339; 319; 393; 309 from least to greatest.

   _____

   Volume 1, Lesson 1-4

Objective: Maintain concepts and skills.

Topic 10: Expressions and Equations

Topic Introduction

Complete with teacher help if needed.

1. The operation modeled below includes an unknown.

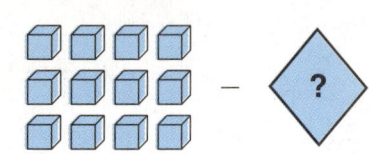

   a. What operation is modeled?

   _____

   b. Which value is known?

   _____

   c. What represents the unknown?

   _____

   Objective 10.2: Write expressions for situations that include an unknown quantity.

2. Evaluate each expression.

   a. $x \div 3$ for $x = 6$

   _____

   b. $8(b - 1)$ for $b = 3$

   _____

   c. $2x + 3$ for $x = 10$

   _____

   Objective 10.2: Write expressions for situations that include an unknown quantity.

3. Calendars cost $10 each.

   a. How would you find the price of 7 calendars? Write an expression.

   _____

   b. How would you find the price of c calendars? Write an expression.

   _____

   Objective 10.1: Record the rule for a pattern as an expression.

4. Kelsey has two boxes of game tokens. One box has 30 tokens in it. Kelsey is not sure how many tokens are in the other box. She knows that she has a total of 75 tokens.

   a. Which values does she know?

   _____

   b. What is the total? _____

   c. Write an addition equation to represent this situation. Use a letter for the unknown amount.

   _____

   Objective 10.4: Solve simple equations for word problems that include an unknown quantity.

Volume 3 — 70 — Level F

Lesson 10-1 — Rules for Patterns

Model It — A

Words to Know   An **unknown** is a letter or a symbol that stands for a number.
An **algebraic expression** is a mathematical phrase with at least one unknown and one operation.

Activity 1

■ + ○○○

The expression is ■ + **3**.

What is the value if ■ = 2?

3 + _____ = _____

Practice 1

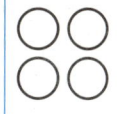

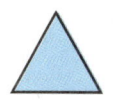

○○ ○○ △

Write the expression.

_____

What is the value if ▲ = 5?

4 + _____ = _____

Activity 2

Hector earns $20 for each hour that he works. How much will he earn next week?

His earnings can be found by multiplying his hourly wage by the number of hours he works.

For **1** hour, he earns 20 × 1 or 20(**1**).

For **3** hours, he earns 20 × 3 or 20(**3**).

For h hours, he earns 20 × h or 20(h) or 20h.

Hector will earn 20h$ next week.

Practice 2

A taco sells for $3. Write an expression to show the price for each purchase.

4 tacos _____

10 tacos _____

300 tacos _____

Write an algebraic expression to show the price for t tacos.

_____

On Your Own

There are 8 black hats and some red hats in a window display. Write an expression to show the number of hats in the display. Find the value if there are 6 red hats.

_____

Write About It

Bags of popcorn cost $2 each. Find the cost of 2, 3, and 4 bags of popcorn. Then write an expression to show the price for p bags of popcorn.

Objective 10.1: Record the rule for a pattern as an expression.

Volume 3   Level F

Lesson 10-1: Rules for Patterns

Understand It

Example 1

Each fruit salad includes two cherries. Colin made a table to show the number of cherries in different numbers of salads. The Input is the number of salads, and the Output is the number of cherries.

Input	2	3	4	5	★
Output	4	6	8	10	2★

The rule for the table is *Multiply by 2*. Another way to write the rule is to use an unknown: 2 × ★ or 2★.

Practice 1

Complete the table to find the cost of different numbers of carnival tickets. The Input is the number of tickets, and the Output is the cost of those tickets in dollars.

Input	5	6	7	8
Output	15	18	21	

Write the rule in words. Then write the rule using an unknown.

_____

Example 2

There are 12 inches in a foot.

Feet	1	2	3	F
Inches	12	24	36	12f

The rule for the table is *Multiply by 12*. The last column of the table shows the rule as an expression: If there are f feet, there are $12f$ inches.

Practice 2

The perimeter of a square is 4 times the length of one of its sides.

Fill in the blanks in the table.

Square's Side Length	1	2	3	4	S
Perimeter	4	8	12		

On Your Own

There are 10 dimes in a dollar.

Fill in the blanks in the table.

Dollar	1	2	3	4	5
Dimes	10	20	30		

Write About It

Look at the table in On Your Own. Write the rule in words. Then write the rule using an unknown.

Objective 10.1: Record the rule for a pattern as an expression.

Lesson 10-1 | **Rules for Patterns** — Try It

1. Write an algebraic expression for each of the following.

   a. $10 + x$ _____

   b. $y - 14$ _____

2. For each table, write the rule in words and as an algebraic expression. Use n for the unknown. Complete each table.

   a.
Input	1	2	3	4
Output	5	10	15	

   Rule: _____

   b.
Input	10	20	30	
Output	16	26	36	46

   Rule: _____

3. Magazines sell for $4 each. Write an expression to show the price for each purchase.

   2 magazines _____

   4 magazines _____

   m magazines _____

4. There were 10 crows in a tree. Then some of them flew away. Write an algebraic expression to show the number of crows that are left. _____

5. $5 + \bigcirc$

   If $\bigcirc = 8$, what is the value? _____

6. Fill in the blanks in the table.

Triangles	1	2	3	4	t
Sides	3	6			

Objective 10.1: Record the rule for a pattern as an expression.

Volume 3 — 73 — Level F

Lesson 10-2 — Write Expressions

A Model It

Activity 1

At the entrance to a shopping center, the sign for the furniture store is twice as long as the sign for the card shop.

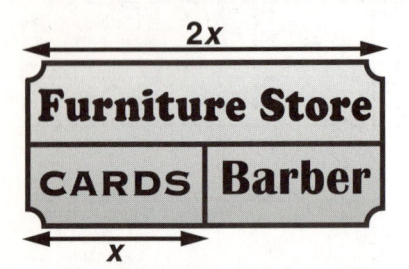

[Diagram: Furniture Store sign labeled 2x above; CARDS sign labeled x below; Barber sign next to CARDS]

The card shop sign is x feet long. The furniture store sign is twice as long, or $2x$ ft.

Practice 1

Complete the picture. Show that Mee's walk to school is four times as long as Cal's walk.

[Cal box shown]

Write expressions for the length of Cal's walk and the length of Mee's walk. Use the same variable in each expression.

Cal _____

Mee _____

Activity 2

The expression shows the cost of x cartons of oil paints.

[Image of carton labeled "Oil Paints $25/carton"]

$25x$

How much would 10 cartons of oil paints cost? Substitute 10 for x:

$25(10) = 250$

The paints would cost $250.

Practice 2

Ms. Archer charges $30 per hour for guitar lessons. Write an algebraic expression for the income from guitar lessons.

hours of guitar lessons: _____

total income: _____

What is Ms. Archer's income if she gives 5 hours of guitar lessons? Evaluate your expression to solve.

_____

On Your Own

A package of hot dogs costs $3. Write an expression for the cost of h packages of hot dogs. Jim buys 3 packages of hot dogs for a cookout. How much does he spend?

_____

Write About It

Sally wrote 4 less than a number as $4 - x$. Is her expression correct? Explain.

Objective 10.2: Write expressions for situations that include an unknown quantity.

Lesson 10-2: Write Expressions

B Understand It

Words to Know
3 fewer than a number $x - 3$
5 greater than a number $y + 5$

Example 1

Francine bought T-shirts for her club. She bought 3 fewer white shirts than black shirts.

If there are x black shirts, then there are $x - 3$ white shirts.

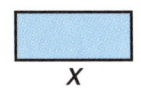

If Francine bought 7 black shirts, how many white shirts did she buy?

Substitute 7 for x: $7 - 3 = 4$.

She bought 4 white shirts.

Practice 1

Amanda's apartment number is 12 greater than Stuart's apartment number.

Write an algebraic expression for each number. Use the variable n in each expression.

Stuart's apartment number _____

Amanda's apartment number _____

What is Amanda's apartment number if Stuart lives in apartment 26? _____

Example 2

Write the algebraic expressions.

Word expression	Algebraic expression
8 less than a number b	$b - 8$
the product of a number a and 8	$8a$
20 divided by a number n	$20 \div n$
20 more than a number r	$r + 20$

Practice 2

Write the algebraic expressions.

Word expression	Algebraic expression
5 greater than a number t	_____
a number q divided by 3	_____
the difference of a number k and 1	_____

On Your Own

Evaluate $4m$ for $m = 7$.

_____

Write About It

Write two different word expressions that can be shown by $3 - x$.

Objective 10.2: Write expressions for situations that include an unknown quantity.

Volume 1   75   Level F

Lesson 10-2   **Write Expressions**   Try It

1. The band's T-shirts cost $15. Posters cost $5. Write an expression to represent the cost of each purchase.

   a. 4 shirts _____

   b. 5 shirts and 9 posters
   _____

   c. n shirts and 6 posters
   _____

   d. c shirts and d posters
   _____

2. Tristan has 6 fewer points than Andre in a video game. Write an algebraic expression for the number of points each boy has earned. Use the same variable in each expression.

   a. Andre _____

   b. Tristan _____

   c. Andre has 85 points. How many points does Tristan have? Evaluate your expression to solve. Show your work.
   _____

3. Write an algebraic expression for each of the following. Use n as your variable. Then evaluate each expression for $n = 8$.

   a. the sum of 7 and a number
   _____

   b. the quotient of 16 divided by a number
   _____

4. Wrapping paper costs $3.00 per package. Which expression gives the total cost of p packages of wrapping paper? Circle the letter of the correct answer.

   A $p + 3$ **B** $3 - p$

   C $p \div 3$ **D** $3p$

5. Explain how to evaluate $3d$ for $d = 4$.

6. Write a word expression for each algebraic expression.

   $b + 6$ _____

   $2b$ _____

Objective 10.2: Write expressions for situations that include an unknown quantity.

Lesson 10-3 **Write Equations**

Model It

Words to Know An **equation** is a statement that two quantities are equal. It is written with an equal sign (=).

Activity 1

Write an equation to show the number relationship for this picture.

♥♥♥ ♥♥ + ? = ♥♥♥ ♥♥♥

5 + ? = 6

Practice 1

Write an equation to show the number relationship for this picture.

? − ★★★★★ = ★★★★★★

_____ − _____ = _____

Activity 2

Juan and his two brothers are at the skate park. Then some other friends joined them. Now there are 7 boys at the skate park. Write an equation that could be solved to find the number of friends who showed up.

Juan and his two brothers: 3
Some other friends: f
Total at the park: 7

3 + f = 7

Practice 2

There are 16 girls on the soccer team. Some of them are out sick this week. There are 12 girls at the game. Write an equation that could be solved to find the number of girls out sick.

Girls on the soccer team: _____

Girls out sick: _____

Total at the game: _____

_____

On Your Own

Vijay brought some cans of food for the food drive. Kelly brought 8 cans. Together, they brought 15 cans. Write an equation that could be solved to find the number of cans that Vijay brought.

Write About It

Write a number story that could be shown by the equation 12 − ▲ = 5.

Objective 10.3: Write equations for word problems that include an unknown quantity.

Lesson 10-3: Write Equations

Understand It (B)

Example 1

There are three kids in the Martinson family. They each brought the same number of sandwiches to the family picnic. There were 21 sandwiches in all. Write an equation that could be solved to find the number of sandwiches each kid brought.

$3 \times s = 21$

Practice 1

Each of the students in Ms. Lopez's class brought in 2 of their favorite books to make a class library. There are 48 books in the class library. Write an equation that could be solved to find the number of students in Ms. Lopez's class.

Example 2

Marc baked 20 muffins for his aunts. He gave each of his 4 aunts the same number of muffins. Write an equation that could be solved to find the number of muffins Marc gave each of his aunts.

$20 \div m = 4$

Practice 2

Lydia picked some flowers from her garden. She gave 4 of the flowers to each of her 6 friends. Write an equation that could be solved to find the number of flowers Lydia picked.

On Your Own

Billie and her 4 friends each played in the same number of matches in the tennis tournament. They played in 40 matches all together. Write an equation that could be solved to find the number of matches Billie and her friends played in.

Write About It

Write a number story that could be described by this equation: $42 \div b = 3$.

Objective 10.3: Write equations for word problems that include an unknown quantity.

Lesson 10-3 — Write Equations

Try It — C

1. Write a number sentence to show the number relationship for each picture.

   a. ▢▢▢ + ? = ▢▢▢▢▢▢▢

   _____

   b. ? − △△△△ = △△△

   _____

2. Monica had 6 clients. She added more clients to her list. Now she has 9 clients. Which equation can be used to find the number of clients Monica added to her list? Circle the letter of the correct answer.

   A $6 + c = 9$ **B** $6 - c = 9$

   C $c + 9 = 6$ **D** $6 + 9 = c$

3. John has a collection of 10 autographs. Julie gives him some more autographs for his birthday. Now he has 13 autographs. Write an equation that could be solved to find the number of autographs Julie gave John.

4. Kay had some students in her dance class. Then 3 of them moved away. Now she has 12 students. Write an equation that could be solved to find the number of students that Kay had originally.

5. At Jonah's party, an extra-large pizza was cut into 18 slices. Each person got 2 slices. Write an equation that could be solved to find the number of people at Jonah's party.

6. Each member of the McGregor family brought 2 cakes to the community bake sale. There were 10 cakes all together. Write an equation that could be solved to find the number of people in the McGregor family.

7. Donatella volunteered to deliver 12 lunches to senior citizens. She delivered some. Now she has 5 lunches left. Which equation could be used to find how many lunches Donatella has already delivered? Circle the letter of the correct answer.

   A $5 + 12 = ?$ **B** $12 - ? = 5$

   C $? - 5 = 12$ **D** $? - 12 = 5$

8. Write a number story that could go with this equation: $y \times 3 = 21$.

Objective 10.3: Write equations for word problems that include an unknown quantity.

Lesson 10-4 — Solve Equations

Model It — A

Words to Know

An **equation** uses the equal sign to show that two expressions have the same value.

The **solution** to an equation is the value for the unknown that makes the equation a true statement.

Activity 1

$$y - 4 = 11$$

Think: what number less 4 equals 11? Since you must take 4 away from the value of y to get 11, y must be 4 more than 11.

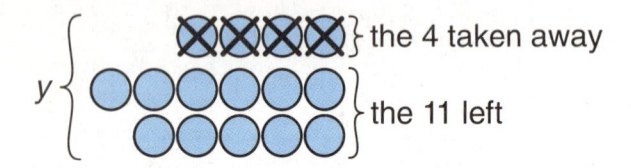

y { ✗✗✗✗ } the 4 taken away
{ ○○○○○○○○○○○ } the 11 left

When 4 were taken away, there were 11 left, so there were 15 to begin with: $y = 15$.

Practice 1

Draw a picture to show $n - 3 = 14$.

What is the value of n?

_____

Activity 2

One box is covered. The total number of baseballs in both boxes is 10. How many baseballs are in the second box?

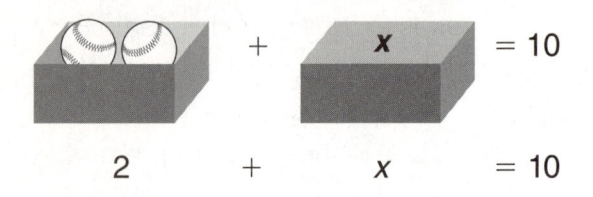

[baseballs] + [x box] = 10

$$2 + x = 10$$

Think: what number added to 2 equals 10? The sum of 2 and 8 is 10, so there must be 8 baseballs in the covered box: $x = 8$.

Practice 2

Two boxes of golf balls are shown below. One is covered. The total number of golf balls in both boxes is 9.

[golf balls] + [x box] = 9

How many balls are in the open box? _____

How many golf balls must be in the closed box if the total is 9? _____

On Your Own

Draw a picture to help you find the solution to $z + 3 = 8$.

_____

Write About It

Use words to write the relationship shown in the following equation: $c - 5 = 15$. What is the value of c?

Objective 10.4: Solve simple equations for word problems that include an unknown quantity.

Volume 3 — 80 — Level F

Lesson 10-4: Solve Equations

Understand It — B

Example 1

Alison had 10 photographs stored in her camera. After the class party, she had a total of 40 photographs stored. The equation below represents this story.

$$10 + ? = 40$$

To find the number of photographs Alison took at the party, use subtraction.

$$40 - 10 = 30$$

Alison took 30 photographs at the party.

Practice 1

Chato spent 30 minutes raking leaves in the front yard. Then he raked the back yard. After a total of 50 minutes, Chato was done with his chore. Write an addition equation to represent this story. Use a question mark to show the missing value.

_____

Find the number of minutes that Chato spent raking leaves in the back yard.

_____

Example 2

Before they opened, Darrell's Bakery baked some doughnuts. By 9 A.M., they had sold 85 doughnuts. There were 35 doughnuts left. The equation below represents this story.

$$d - 85 = 35$$

To find the number of doughnuts Darrell's Bakery baked, use addition.

$$35 + 85 = 120$$

Darrell's Bakery baked 120 doughnuts.

Practice 2

Mr. Powell had some students in his beginner swim class. 40 of them graduated to the advanced class. 55 remained in the beginner class. Write a subtraction equation to represent this story.

_____

Find the number of students who started out in the beginner swim class.

_____

On Your Own

There are two museums in town, the Fine Art Museum and the Modern Art Museum. There are 320 paintings at the Modern Art Museum. All together, they have 610 paintings. Write an addition equation, and find the number of paintings at the Fine Art Museum.

Write About It

Write a number story that could be represented by this equation. Then find the value for x.

$$95 - x = 12$$

Objective 10.4: Solve simple equations for word problems that include an unknown quantity.

Lesson 10-4: Solve Equations

Try It

1. Use mental math to find each missing value.

   a. ? − 20 = 50 _____

   b. ? + 15 = 25 _____

   c. 14 − ? = 6 _____

2. Use mental math or draw a picture to find each solution.

   a. 16 − e = 4 _____

   b. f + 9 = 13 _____

   c. 10 + g = 42 _____

   d. j − 11 = 3 _____

3. The park district plans to have 10 soccer teams this summer. They will also have some volleyball teams. All together, there will be 24 volleyball and soccer teams. Write an addition equation to represent this number story. Then find the number of volleyball teams.

   _____

4. Charlie had 32 music students. Some of them studied guitar. 24 studied piano. Write a subtraction equation to represent this number story. Then find the number of students who studied guitar.

5. Write a number story that could be represented by this equation. Then find the value for y.

   y + 45 = 60

6. Omar had $13. His friend loans him some money. Now Omar has $18. Write an addition number relationship using a ? mark to show how many dollars Omar's friend loaned him. Then solve.

Objective 10.4: Solve simple equations for word problems that include an unknown quantity.

Topic 10: Expressions and Equations

Topic Summary

Choose the correct answer. Explain how you decided.

1. Jill owns a pool-cleaning service. She had 8 pools on her schedule today. She has 3 pools left to clean. Which is **not** a way to represent this number relationship?

   A $8 - x = 3$

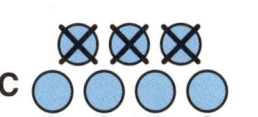

   B [8 circles] $- x =$ [3 circles]

   C [array of circles with 3 crossed out]

   D $x +$ [3 circles] $=$ [8 circles]

   _____
   _____
   _____

2. Jack's Audio and Video is having a big sale. All CDs will sell for the same price. The price of a DVD will be $5 more than the price of a CD. If the price of a CD is c dollars, which expression represents the cost of a DVD?

   A c

   B $5c$

   C $c + 5$

   D $c - 5$

   _____
   _____
   _____

Objective: Review writing and solving expressions and equations.

Topic 10: Expressions and Equations

Mixed Review

1. Compare using < or >.

a. 180 ◯ 108

b. 86 ◯ 231

c. 809 ◯ 908

d. 131 ◯ 126

Volume 1, Lesson 1-4

2. Find each product.

a. 302 × 5 = _____

b. 2,316 × 4 = _____

c. 2,040 × 3 = _____

d. 538 × 1 = _____

Volume 3, Lesson 7-3

3. A hot dog stand sells 367 hot dogs a day. If the cost of a hot dog is $3.15, about how much does the hot dog stand make in one day? Circle the letter of the correct answer.

A $900   B $1156.05

C $1,200   D $1,600

Volume 3, Lesson 7-2

4. Which shows six million, three hundred five thousand, nineteen in standard form? Circle the letter of the correct answer.

A 635,019   B 6,350,019

C 6,305,019   D 60,035,190

Volume 1, Lesson 2-3

5. Find each sum. Show how you added.

a. 6 + 2 + 7 = _____

b. 3 + 9 + 4 = _____

c. 1 + 5 + 7 = _____

d. 8 + 6 + 9 = _____

Volume 3, Lesson 9-1

6. Find the value of each expression for the given value of the variable.

a. $a = 5$; $a + 2$ _____

b. $b = 2$; $6b$ _____

c. $c = 7$; $c - 4$ _____

d. $e = 9$; $e \div 3$ _____

Volume 3, Lesson 10-3

7. Write a related subtraction fact for 8 + 3 = 11.

_____

Volume 2, Lesson 3-4

8. Write 3 numbers that are divisible by both 5 and 10.

Volume 2, Lesson 5-3

Objective: Maintain concepts and skills.

Words to Know/Glossary

A

algebraic expression — A mathematical phrase with at least one unknown and one operation.

associative property of addition — Group numbers in any way and the sum will be the same.
For example, $(2 + 8) + 7 = 17$ and $2 + (8 + 7) = 17$.

associative property of multiplication — Group numbers in any way and the product will still be the same.
For example, $(2 \times 7) \times 8 = 112$ and $2 \times (7 \times 8) = 112$.

C

commutative property of addition — Add two numbers in any order and the sum will be the same.
For example, $6 + 5 = 11$ and $5 + 6 = 11$.

commutative property of multiplication — Multiply two numbers in any order and the product will be the same.
For example, $5 \times 6 = 30$ and $6 \times 5 = 30$.

compatible numbers — Numbers that are close to the actual number but easier to work with mentally.

D

distributive property —Multiplying a sum (or difference) by a number gives the same result as multiplying each number in the sum (or difference) by the number and adding (or subtracting) the products.
For example, $8 \times (2 + 7) = (8 \times 2) + (8 \times 7)$.

E

equation — A statement that two quantities are equal. It is written with an equal sign (=) to show that two expressions have the same value.

estimate — An estimate is a number close to the actual number.

M

mental math — Solve math problems in your head.

multiple —The product of a given number and another whole number.

R

regroup — To exchange amounts of equal value to rename a number.

remainder — An amount left over when one number is divided by another.

round — To change the digits in a number to make a similar but easier to use number.

S

solution — The solution to an equation is the value for the unknown that makes the equation a true statement.

U

unknown — A letter or a symbol that stands for a number.

Word Bank

Word   **My Definition**   **My Notes**

Word **My Definition** **My Notes**

Index

A
about, 36, 53–54
addition
    estimation with, 8–11
    of multidigit numbers, 20–22
    of 1-digit and 2-digit numbers, 2–4
    of 2-digit numbers, 11–13
    of 3-digit numbers, 17–19
    properties of, 59–61
    using to compute, 1
    using mental math, 5–7
algebra, basic properties of, 58
algebraic expression, 71
approximately, 36, 50, 54
associative property
    of addition, 59–61
    of multiplication, 62–64

C
commutative property
    of addition, 59–61
    of multiplication, 62–64
compatible numbers, 45

D
distributive property, 65–67
dividend, 41
division
    choosing a method for, 50–52
    multidigit numbers by 1-digit numbers, 47–49
    multiples of 10, 100, and 1,000, 41–43
    problems, 53–55
    using to compute, 40–42
divisor, 41

E
equal sign (=), 77
equations
    definition of, 72, 77
    and expressions, 70
    solving with unknowns, 80–82
    writing with unknowns, 77–79
estimate, 8, 29, 44
    products, 29
    quotients, 44–46
expressions
    algebraic, 71
    and equations, 70
    writing, 74–76
    writing for pattern rules, 71–73

F
fewer, 75

G
greater, 75

M
mental math, 5–7, 35, 44–46
Mixed Review
    Topic 6: Use of Addition or Subtraction to Compute, 24
    Topic 7: Use Multiplication to Compute, 39
    Topic 8: Use Division to Compute, 57
    Topic 9: Basic Properties of Algebra, 69
    Topic 10: Expressions and Equations, 84
multiple, 26
multiplication
    choosing a method for, 35–37
    multidigit and 1-digit numbers, 32–34
    multiples of 10, 100, and 1,000, 26–28
    problems, 53–55
    properties of, 62–64
    using to compute, 25

N
number line, 8

P
products, estimating, 29–31
properties
    of addition, 59–61
    of algebra, 58
    of multiplication, 62–64
    See associative property;
commutative
        property; distributive property

Q
quotient, 41
    estimating, 44-46

R
regrouping, 14
remainder, 47
rounding, 29–31, 44–46

S
solution, 72
subtraction
    estimation with, 8–11
    of multidigit numbers, 20–22
    of 1-digit and 2-digit numbers, 2–4
    of 2-digit numbers, 14–16
    of 3-digit numbers, 17–19
    using to compute, 1
    using mental math, 5–7

T
Topic Summary
    Topic 6: Use of Addition or Subtraction to Compute, 23
    Topic 7: Use Multiplication to Compute, 38
    Topic 8: Use Division to Compute, 56
    Topic 9: Basic Properties of Algebra, 68
    Topic 10: Expressions and Equations, 83

U
unknowns, 71, 74–82
    solving equations with, 80–82
    writing equations with, 77–79

W
writing
    equations with unknowns, 77–79
    expressions, 74–76
    expressions for pattern rules, 71–73